T̶ ̶ ̶:t
Experience

The Way of Non-Duality

Alessandro Sanna, PhD

Copyright © 2023 Alessandro Sanna
All rights reserved.
ASIN: B0C8FSNPSX (ebook)
ISBN: 979-8393753498 (paperback)
ISBN: 979-8857866405 (hardcover)

No part of this book may be reproduced in any form or by any means, electronic or mechanical, including photography, recording, or any information storage and retrieval system or technology now known or hereafter developed, without written permission of the publisher.

To contact the author, you can send an email to:
dr.alessandrosanna@gmail.com

First edition: June 2023
Second edition: November 2023
A project by: 2Samaræ

Cover photo by: Aina S. Erice

Dedicated to Aina, Leire Gaia, Maurizia and Zio Renato

All that can be perceived is not the perceiver.
Huang Po (Chan/Zen Buddhism Master)

The world is given to me only once, not one existing and one perceived. Subject and object are one.
Erwin Schrödinger

If you can think of it, then it's not you: the conscious subject is separate from the known object.
If it is not you, then you cannot think it: the known object is not separate from the aware subject.
Alessandro Sanna

God created the senses to face outward. Therefore, one sees the external world, no one's own nature. A wise man, desiring to be immortal, looks within and sees the true Self.
Katha Upanishads 2.1.1

Alessandro Sanna					The Direct Experience

INDEX

Index	7
Preface	9
1. Introduction	13
2. Materialism	17
3. Beyond Materialism	26
4. Consciousness	33
5. Argument From Quality	44
6. Argument From Simulation	53
7. Argument From Representation	59
8. Argument From Asymmetry	64
9. Argument From Becoming	75
10. Argument From Substantiality	79
11. Argument From Entropy	84
12. Argument From Presence	88
13. Argument From Ineffability	100
14. Argument From Concepts	112
15. The Meaning of Experiences	116
16. The Multiplicity of Experiences	121
17. The Triangulation of Experiences	125

18. Epistemology of Direct Experience — 131
19. The Direct Experience: The Negative Path — 135
20. The Direct Experience: The Positive Path — 144
21. The Experience of Meaning — 150
22. Conclusion — 159
Appendix. Philosophical Principles — 162
Biography — 165
Acknowledgment — 167
Bibliography — 169

Preface

Talking of the 'witness' should not lead to the idea that there is a witness and something else apart from him that he is witnessing. The 'witness' really means the light that illumines the seer, the seen and the process of seeing. Before, during and after the triads of seer, seen and seeing, the illumination exists. It alone exists always.

Sri Ramana Maharshi

The deepest purpose of life is to discover who or what we truly are.

Perennial Philosophy

Q: What is Consciousness?

A: Look around you... are you aware of what you see?

Q: Yes

A: Consciousness is the name commonly given to that which is aware of what you see, hear, think, etc. It is also called Awareness.

Q: And what relationship does Consciousness have with me?

A: It has an identity relationship. You are Consciousness, Awareness. But not only that: you are solely and exclusively Awareness. During dreams you change your body, your personality, your memories, you change many things, but you always remain the same Awareness. You have a body and a mind, but you are not a body and a mind; you are Awareness.

Q: And what is the relationship between Consciousness and the external world?

A: Where is the outside world located? Where are concrete things like the floor or clouds found? Where do you find abstract things like the monetary system or the immune system?

Q: Somehow they're out there.

A: No, pay more attention, try to investigate more deeply: what does your direct experience show you? It shows you that everything you know, that is, everything you come into contact with, is found within your Consciousness.

Q: Yes, let's say it's as if inside my head there was a homunculus looking at the outside world as if he were sitting in an opera house watching an orchestra.

A: No, it's not like that. Be more careful, it's not like watching an orchestra sitting inside an opera house. It is rather like being the opera house itself: which contains within itself all the reality of the concert.

Q: However the concert inside my consciousness is a representation of a real concert taking place out there in the physical world.

A: There is no representation, rather there is a presentation, that is, a manifestation of consciousness to itself and in itself in the form of the various known objects. Let me explain a little better. Consider a video surveillance system. A camera records what happens in a room and a computer analyzes the footage. In this example we clearly distinguish the known object, the knowing subject and the activity of knowing. The known object is the film. The computer is the knowing subject. Finally, recording the film is the activity of knowing. The cognitive process of the video surveillance system is a triad composed of three elements: knowing subject, known object and cognitive activity. These 3 elements are clearly distinct and separate from each other. More generally we can state that in any physical reality in which a cognitive process can be found, the cognitive triad is always present. The fact that there is this triad: that is, that object, subject and cognitive activity are 3 separate things, immediately implies that there is a reality existing outside the subject, a reality to which the subject has access. Are you following me?

Q: Everything is quite clear.

A: Conversely, in your direct experience, when you know anything the triad collapses into a monad: subject, object and cognitive activity are a single indivisible substance. Check for yourself right now. Try, for example, hearing a sound. Realize that you are not at all able to separate between them: hearing (cognitive activity), sound (known object), and the awareness you have of the sound/hearing (the subject). At this point, and unlike what happens in a physical process, it appears that the subject-awareness in the act of knowing an object is not accessing a

1. Preface

reality external to itself, because the subject and object coincide, they are not separate. All reality is a manifestation of and in consciousness.

Q: Wow, what you say is so clear, simple and obvious. Yet I had never paid the slightest attention to it. This changes everything. But in the same way I also realize that the prudence of my skeptical and scientific mentality is already giving rise to a cascade of questions, doubts and possible objections. Would you be able to develop this topic much more in depth and possibly with both an analytical and experiential approach?

A: Naturally, in this book you will find the 10 most convincing analytical arguments in favor of the primacy of consciousness over matter, accompanied by experiential considerations. I'm sure they will leave you completely convinced. Read to believe.

Who am I? What is the nature of reality? Why is there something rather than nothing? What is the meaning of existence?
The "non-dual" understanding of reality provides enlightening and reasonable answers to all these perennial questions. "Non-duality" is based on the idea of the primacy of consciousness over matter: matter is a manifestation of consciousness, and not vice versa. Non-duality was defined by Aldous Huxley as the "perennial philosophy", because it has been discovered, or glimpsed, repeatedly since ancient times in almost all cultures and historical eras.

This book provides a surprising initiation into the perennial philosophy of non-duality through the perspective and arguments of a computational physicist. The result is particularly understandable, enlightening and encouraging.

In this book the author has developed and presents 10 suggestive, convincing and compelling analytical arguments in favor of non-duality. These arguments are a game changer.

Some of the things you will find in this book:
- A profound analysis of materialism;
- Why materialism is incomplete;
- The way to go beyond materialism;
- A precise description of the non-dual understanding of reality;

- 10 analytical arguments in favor of the primacy of consciousness over matter;
- Answers to all possible counter arguments;
- Instructions for having an experiential understanding of non-duality;
- What can we expect after the death of the body;
- And much more ...

Example of some of the benefits of reading this book:
- Radical change of perspective on reality: we move from identifying ourself with our own body and/or mind, to understanding experientially that we completely transcend this physical plane, with all the implications that this entails.
- Be able to justify the primacy of consciousness over matter before oneself and everyone else, through both rational and experiential arguments.
- Discovery that the qualities of experiences cannot be reduced to empty abstractions devoid of feeling: there is meaning in every experience.
- Discovery of the existence of a profound inner dimension of peace, underlying any possible pleasant or unpleasant experience;
- And much more ...

Reading this book will radically change your vision of reality for the better.

1.

INTRODUCTION

Problems that always remain unsolved should be suspected of being questions asked the wrong way.

Alan Watts

Fools reject what they see, not what they think.
The wise reject what they think, not what they see.

Huang Po (Chan/Zen Buddhism Master)

The truth is neither one nor two. It is as it is.

Sri Ramana Maharshi

Most of us grew up and live in a society where the materialist-mechanist paradigm was established as truthful and scientific. In this paradigm, we and everything that exists can only be characterized in terms of physical entities that, interacting appropriately, produce everything we find in reality.

According to this school of thought, any phenomenon can be explained with two kind of causes: the parts that compose it plus appropriate rules or laws of interaction between the parts. By iterating this procedure in a fractal way on the parts of the parts, one goes in search of the components and of the fundamental laws that can explain all of reality in a totally mechanical way.

In this paradigm, qualitative phenomena such as feelings are reduced to chemical reactions in the brain; laughter is reduced to its physiological aspect only.

In this paradigm, the universe appears hostile, boring, meaningless, aimless, and dark. Life is ultimately short and ugly, and the best you can do is try to capture as many pleasant experiences as possible, knowing that they are only anesthetics to the sense of nihilism.

As Steven Weinberg said perfectly, the gist of the discourse is that: "The more the universe seems understandable, the more it also seems useless".

However, a careful observation of any human culture that has ever existed reveals that our deepest intuitions and practical behaviors diverge totally from this paradigm.

Many of us today are either explicit materialists or are at least implicit materialists. The explicit are those who have made peace with reality and who have ceased to deceive themselves and openly profess a materialistic worldview.

The implicit are those who have not made peace with reality, but who "know how things really are", and who behave adequately about it, but who nevertheless keep deep inside themselves the hope, or the illusion, that there could be a different, inspiring and valid way of how to see reality, even if they haven't quite figured out how yet.

Today, many Christian theologians are also implicit materialists, and have developed theologies in which people are no longer endowed with an immortal soul, as everything probably ends in this life.

The materialist view is well constructed and intellectually attractive. It is like classical religions: it is a complete and all-encompassing worldview that leaves no gaps.

The best argument in favor of materialism can boil down to the following: "we have washing machines, therefore materialism is correct". Materialism seems to have brought so many advantages to humanity that no alternative is anywhere near its equal.

But is it really so? I suggest in this book that this is not the case, but instead, that we have good reason to think that materiality is not fundamental. That, at the ground of reality, there are no inert entities but that the exact opposite occurs.

1. Introduction

Is it possible to argue against materialism without making reason to commit suicide? I really think so. In this essay, I aim to argue that materialism is not reasonable and I will illustrate an alternative.

The alternative comes from placing our attention on a mystery that we constantly have before our eyes: awareness. It is said that the best way to hide something is not to hide it at all. Because we are led to go in search of hidden or semi-hidden things, neglecting what we constantly have under our noses. In this book, I will show that focusing on awareness we can amazingly clarify many issues.

In the next chapter, I will explain the reasons for the materialist paradigm trying to show its conceptual strengths and the charm it can have.

In the third chapter I will show the weaknesses of materialism, and then, chapter after chapter, I will build the argument for the non-dualism paradigm.

This book does not intend to be exhaustive at all, but only wants to introduce or rediscover a different way of thinking, which in philosophy is called non-dual understanding. Non-dualism was defined by Aldous Huxley as the "perennial philosophy", because it is an idea that has arisen in all ages and in the most disparate cultures. We recognize it in the West in some Christian mystics, in Sufism and in the Kabala; in many philosophical paradigms such as Neoplatonism, or some forms of Idealism, and in the East in the philosophical traditions linked to Hinduism, in particular Advaita Vedanta, in various schools of Mahayana Buddhism, in Taoism and in Zen.

What is said in this essay is therefore nothing new, and nothing exhaustive. Rather, I will try to present perennial ideas in the way I would have wanted them explained to me when I started to approach them. Coming from a scientific background, and reading certain kinds of ideas, I get specific kinds of doubts and objections, which probably don't come to people with another kind of training. I also feel the inner need to develop analytical arguments in favor of a certain paradigm before being able to adopt it. Therefore, if whoever is reading these pages recognizes themself in a scientific mindset, know that this book is somehow dedicated to them.

In the end, the acceptance or not of a certain paradigm is more an emotional question than a rational one; more from the heart than from the head. But for researchers and readers like me, reason is the bouncer of the ballroom of the heart. Before you can accept something emotionally, reason must be convinced.

I tried to keep the number of pages as small as possible, so that the shortest possible book came out. This way the potential reader doesn't feel scared by the number of pages and is more likely to start reading it. However, this implies that the pace of the book is quite sustained and dense in content. I suggest reading this book from beginning to end, without jumping and without interruptions. Then it is to put it aside and stay for a certain time reflecting on the content, looking for objections or confirmations, and above all verifying that it reflects one's direct experience. Having done this, I advise you to iterate the procedure: read it and then think about it a second time. Lastly, one should do nididhyasana: non-dual meditation, try to notice experientially the non-duality of reality.

2.

MATERIALISM

Do not multiply objects unnecessarily.

William Occam

Thinking is hard, that's why most people judge.

Carl Gustav Jung

In the Western cultural tradition, the materialist paradigm was officially born with the ideas of the Greek philosopher Democritus. Democritus had imagined that all reality could be explained in terms of indivisible particles, called atoms, which moving, colliding and aggregating in space, without any purpose, gave rise to all the phenomena of the universe.

Democritus' theory was much despised by Plato, who, in one of his dialogues, the Timaeus, criticizes Democritus' theory without even mentioning Democritus: he despised him so much. After Plato came Aristotle who, although not despising Democritus, did not spare him harsh criticism.

What Aristotle particularly lacked in Democritus' system were formal and final causes. We must imagine that after Parmenides the Greek philosophers were busy explaining the possibility of becoming: that is, of change. How is it possible for something that did not exist to begin to exist or to cease to exist? In the Aristotelian system, the becoming, the change, of an entity X is explained in terms of 4 causes:
1. The efficient cause: that is, the agent who pushes X to change;

2. The material cause: what X is made of.
3. The formal cause: what X becomes.
4. The final cause: that is, the directionality between what X was and what X became.

For Aristotle and for much of later philosophy, the necessity of all 4 causes was something very evident.

In Democritus' materialist approach, instead, only the first 2 causes are required: the efficient one and the material one. In this sense, atoms, as a whole, are both the material cause and the efficient cause of reciprocal motion: like billiard balls, but without a cue. The billiard cue is the one that gives directionality (final cause) in the movement of the balls and brings them towards the new configuration (formal cause). But if the cue is not present and the balls move by colliding with each other it is superfluous to add the last two causes. Reality is thus reduced to a mechanics of free bodies in motion: this is where the name of mechanism often given to materialism comes from. In this scheme Aristotle contested Democritus that even if the balls transmitted the motion to each other, there must still always be a first movement which, starting from the completely stationary balls, started the game. That is, there should be at least one initial move of the cue that started the game: a kind of immovable engine (we will see the exact definition of an immovable engine in the next chapters). But Democritus could have replied that the balls have always been in motion, and therefore there is no need for the cue to start the game.

After the criticisms of Plato and Aristotle, Democritus' theory remained largely frowned upon. After which the centuries passed and the last medieval scholasticism arrived which, unfortunately, sank and got lost in its own conceptual depth. Philosophy had reached a dead end and a refoundation was needed.

At that point Descartes arrived to try to re-found philosophy in what we now call modern philosophy. It was the period in which modern science was establishing itself, which, with its empirical and mathematical method, demonstrated that it had a way of asking nature questions so that the latter could do nothing but answer them. Descartes set out to apply the same method to philosophy, so that from now on

2. Materialism

when 2 philosophers had dissenting opinions, there was a precise mathematical methodology to figure out who was right.

Unfortunately, however, the study method must adapt to the object of study and not vice versa. I cannot apply the mathematical method of physics to philosophy. In the same way I cannot apply the study method of psychology to meteorology.

As a result, Descartes' hopes were soon dashed, and again there began to be as many disputes and as many different opinions as there were different philosophers.

But what Descartes managed to keep as his legacy was the almost universal tendency of modern philosophy to look at the physical and natural sciences with a sense of inferiority and imitation. Nowadays many philosophers, especially those of analytical current, are divided between those who write works on the uselessness of philosophy, and those who seem to write treatises on physics-mathematics applied to aerospace engineering, but without any real application.

In this climate, philosophers have generally stopped answering the questions pertaining to philosophy and have thus left uncovered a field that has necessarily gone on to be occupied by other figures. These other figures are often scientists who try to answer purely philosophical questions with the generally inappropriate method of their scientific field.

It is thus that the materialistic-mechanistic method of physics, perfectly suited to physics, has been used to answer philosophical questions. Democritus had his final revenge on Plato and Aristotle.

But exactly, nowadays, in what form has "Democriticism" evolved and presents itself?
I think it can be said that at the basis of the mindset of contemporary Democritism (read: mechanistic-materialism) is the assumption that everything can be explained in computational terms: all of reality is one big computation. While the ancient Democritus saw

atoms moving aimlessly through space, the modern Democritus now sees Turing machines, computers, performing a calculation. A calculation is something that in itself is devoid of choice, emotion, color. A calculation is purely mechanical and analytical, it has no form or purpose.

An emblematic expression of this attitude is the simulation theory according to which our universe is nothing more than the calculation of a mega computer, exactly like a highly sophisticated video game.

But in this video game what are emotions, colors and choices? They are nothing in themselves, because everything that is not computation is just a way we humans have to describe computation efficiently, without too many words. This way of thinking is what in philosophy is called, by some, "eliminativist" and by others "poetic naturalism". Eliminativism means eliminating the hypothesis that there are things that are not subject to calculation; while poetic naturalism means that colors, emotions and choices are just a poetic way of expressing a mathematical equation or an algorithm or a set of data.

For example, let's take the concept of a planet's center of gravity. The rotational movement of a planet on its axis (or around its star) can be described with the equations of motion of each single atom that composes it. Such a description would use an absurdly large amount of information. But there is also a much more compressed way of describing rotational motion. It is enough to introduce the abstract concept of the planet's center of gravity, and describe the motion of this single imaginary point. The movements of all the other points of the planet can be expressed with suitable vector sums with respect to the center of gravity point. The resulting algorithmic compression between these two descriptions is immense. Well, emotions, sensations and all mental qualities are to the very complex neural network of the brain, as the center of gravity is to the planet. They are just names that facilitate description, but they don't really exist in themselves; just as the center of gravity does not physically exist in itself, but only as a mathematical model.

2. Materialism

But how does computationalism, that is, the contemporary form of Democritean materialism, try to answer the classic questions of philosophy? Let's see it now.

Where does the universe come from ?
The universe came out of nowhere, specifically it comes from a random fluctuation in the quantum vacuum, which for no specific reason, especially no man-made reason, eventually led to humans appearing on an outlying planet of an outlying galaxy.

What are the objects in themselves? What is a pizza or a feeling?
An object is a set of relationships with other objects. Everything is explicable in terms of a substrate plus an organizational structure superimposed on the substrate. And what is this substrate? This substrate is in turn a way of organizing a deeper substrate. And so on in a fractal way. A feeling is the pattern of electro-chemical signals that certain neuronal circuits exchange. These circuits are the organization scheme of a group of neurons. And neurons are a pattern of organization of a group of molecules. And molecules are a pattern ... Patterns as such are entities that can be perfectly reproduced in a software: they are computational entities. Therefore everything is a colossal computation.

Is there free will?
No, because the schemes (physical laws or algorithms) are either deterministic (classical physics) or they are indeterministic (quantum physics). If our brain responds to deterministic laws, then our behavior has already been determined. If our brain responds to indeterministic laws then our behavior is not free but random.

Why does the mother love her children?
Because it is an evolutionarily advantageous behavior. In evolutionarily historical times, mothers who had the (genetic) predisposition to love, and therefore to worry about, their children allowed their offspring (who had inherited the same genetic predisposition to care for their children) to have more chances to survive childhood and eventually to have children of their own. But why did a first mother appear in the first instance with the predisposition to take

care of children? By pure random genetic mutation. The feeling of maternal love is nothing but the result of chance and evolutionary convenience; there is no other meaning to be found behind this feeling.

Why are we sociable and some behave well? Where does morality come from?

Because in evolutionary historical times the groups of people who had the genetic predisposition to collaborate, within certain limits, with the members of their group survived the most. In this way the more cohesive groups have defeated the less cohesive and less organized groups, and have therefore had access to more resources (hunting, fishing, breeding, etc..) and therefore have reproduced more. This attitude has led, in general, to the development of intra-group moral and collaborative behaviors, and extra-group hostile and tribal behaviors. For this reason humans are self-righteous and xenophobic.

Why is economic and social development?

Thanks to the invisible hand of the market. Greed has proved to be a great driving force for economic development. Avarice allows the most capable to excel thanks to the search for innovative methods, products and strategies that inevitably end up spreading and benefiting the whole of society. The benevolent secondary effect, that is: the fact the it is meritorious for the fittest, is the formation of hierarchical social classes according to a power law.

What is the truth ?

There is no truth. There are rather post-truths. That is, behaviors and affirmations (called in technical terms "memes") which reveal themselves, over time, evolutionarily more adaptive. Memes follow the same algorithm as genetic evolution. We will therefore have memes that, as viruses of the mind, have more ability to spread to other minds than others. You believe and do what you do because you have particular memes in your head. You find yourself having such memes in your head because they have allowed you more leeway than other minds with other memes. Who decides what a utility is? Another kind of meme. In fact, memes, like genes, don't travel alone from one mind to another, but travel in compact (tribal) tribes to maximize viral chances. The

2. Materialism

"monotheistic God" meme travels well with the "if you don't believe in such a God you go to hell" meme. In fact, religions that instill fear of hell tend to be more difficult to eradicate, so much so that even Blaise Pascal fell into this trap by rationalizing it with his famous wager. Minds, like organisms for genes, are machines used by memes to reproduce; nothing more.

How do you judge behavior?
Based on the relative utility margin compared to another behavior: utilitarian ethics.

Why is there a universe like ours that has apparently so well calibrated parameters for the birth and evolution of life?
Simply because there are an indefinite (or infinite?) number of parallel universes in which the fundamental physical constants are not adequate for the birth and development of life. It is fair that in retrospect we can say that we are in one of the few universes appropriate for life. There is nothing purposive about any of this.

Is there any meaning in life?
Meaning doesn't exist. The theory of evolution has shown that meaning is an illusion. Natural selection is a "forward" mechanism that behaves as if it were a retroactive (that is: finalistic) mechanism. In a retroactive mechanism, a process compares the result it obtains with a desired objective (the goal) and if there is a difference, then it acts appropriately on the efficient causes upstream of the process so as to reduce the difference downstream. The retroactive mechanism has a goal.

In natural selection, on the other hand, there is a population of processes, each more or less different from the other. However, these processes share the use of some limited resources. Some processes will be able to grab all available resources; while others will not have enough to complete. The latter will be discarded; while the former will eventually be able to combine with each other to generate a new generation of processes that replace the discarded part of the population. After a certain number of iterations, we arrive at a population of processes optimized to obtain a certain result, without having sought this result

23

intentionally. There is therefore no intentionality in nature; nothing was desired or wanted for itself: neither you, nor me, nor anything else.

Why do I exist?

Because a blind mechanism produced you at a certain moment, without any specific interest for you. You are just one of the many organisms of a certain population subject to the mechanism of evolution. A single organism has no value in itself. What has value, evolutionarily speaking, is the population.

Who am I?

You can be defined in several ways: your body; your brain; the neural network of your brain; a tool used by genes and memes to reproduce; a consumer within the capitalist system; a victim and/or perpetrator of the social power system; the food you eat; your social role; etc …

Yes, fine, but what is my essence? Am I not conscious?

There are no essences of things, and consciousness is an illusion or hallucination produced by the brain. Consciousness is like an optical illusion, it is a way of speaking. In truth you are not conscious. The seat of consciousness in the brain has never been measured or found. Therefore consciousness does not exist.

How do we relate to the Amazon and the environment in general?

There is no proper ethics. Life is short and ugly. The only thing we can try to do is to maximize the pleasant moments. Everything else bends to this need. So the fate of the Amazon and the environment depends on their usefulness for this purpose. Will I get rich if I cut down the Amazon for wood and palm oil plantations? Perfect then let's take out the Amazon.

But are all these answers, albeit a little depressing, really correct?

Obviously they are, and the definitive proof comes from the tremendous success that science and technology have had in understanding the world and bending it to our needs. Never in the

2. Materialism

history of humanity have we been so good as now. Ultimately: we have the washing machine ergo materialism is correct.

3.

BEYOND MATERIALISM

Materialism is the philosophy of the subject who forgets to take into account itself.

Schopenhauer

In theory there is no difference between theory and practice, while in practice there is.

Albert Einstein

I will not commit the fashionable stupidity of considering everything I cannot explain like a fraud.

Carl Gustav Jung

The materialist proposal is suggestive enough, but is it also the definitive proposal?

I think it is by no means the final proposal; I think it can be said that what materialism proposes is to some extent correct but incomplete. The answer to materialism is not a "No", but a "Yes and also".

Let us therefore try to complete the responses of materialism seen in the previous chapter, using a broader perspective.

Where does the universe come from ? ...

The nothingness that physicists talk about is not nothingness in the philosophical sense. In a philosophical sense, nothingness is neither an actuality nor a potentiality. It is not even imaginable, it is not really cogitable. This is why it is defined in negative and not positive terms.

3. Beyond Materialism

The word "nothing" is by definition devoid of external reference, it is a meaningless signifier, or in computer terms: an unassigned pointer. On the contrary, the nothingness of physicists is the quantum vacuum, which is defined in positive terms, that is, with affirmative sentences: "the quantum vacuum is a field of probability…". So the quantum vacuum is something. By definition nothing is not something. The materialist would answer: yes, but if time was born with the big bang, the question of what was before doesn't even make sense: if time came into existence with the big bang then there is no need for an immobile mover that started games.

Well, the idea of the immobile mover is one of the most misunderstood ideas of classical philosophy in modern philosophy and science. Classical philosophy called the immobile mover not the first of the efficient causes along a temporal sequence; but rather the first of the hierarchical causes. The immobile mover (unlike the concept of the modern Demiurge) does not cause the universe in time like the billiard cue causes the first ball to move and then crash into the other balls. The immobile mover is more like the pool table that supports the game (all the balls and their movements) before, during and after the game. The terms "before, during and after" are to be understood in an ontological and not a chronological sense.

What are the objects themselves? What is a pizza or a feeling? …

It's fine to use patterns to be able to understand and manipulate the world, but let's be careful not to confuse representation with presentation. A pattern is a representation inside our mind of something that's out there outside our mind. A representation is not the object represented, but is only a thought. The painter Magritte captured this fact well with his famous painting entitled: "The Betrayal of Images"; where a pipe is drawn with the caption: "This is not a pipe". In the sense that a drawing of a pipe is not a pipe. The drawing of the pipe represents a pipe, but in itself it is something else. The legendary mathematician John Von Neumann once said that: "The sciences don't try to explain, they almost don't even try to interpret, they mainly make models. By model we mean a mathematical construct which, with the addition of some verbal interpretations, describes the observed phenomena. The justification for such a mathematical construct lies solely and precisely in the fact that it is expected to work".

A representation always refers to something external to our mind, otherwise we will fall into solipsism. A feeling, a pizza, or anything whatsoever is not the representation, the model, the scheme that we make of it in our mind, but they have their own reality.

Is there free will? ...

To question the existence of free will, we don't even need to reason about the fact that physical laws are deterministic or nondeterministic, or remind ourselves of Lisbet's experiments. It is enough for us to do the following experiment: let's try to predict our next thought. Obviously we can't do it because what we would have predicted will be the second thought at best. If we could choose our thoughts then we should never be unhappy, because we will always choose to feel happy feelings, happy images, and happy words. But if we fail to predict our next thought then we are not in command of our mind, and therefore we are not free.

All clear enough, I suppose. But then one can ask oneself the meta problem. If I'm not free, then my reasoning, my argument is not free either. That is: my reasoning, and in general what I think (whatever I think), is not impartial: it is partly determined in a deterministic way by the initial conditions at the moment of the big bang, and in part it is determined in an nondeterministic way by random quantum fluctuations. It is therefore not determined by the validity or otherwise of its intrinsic conceptual content, by the truth value of an affirmation. Therefore the "believing that we are not free" does not depend on the fact that actually "we really are not free".

But then are we free or not? When paradoxes like this are presented to us, there is a clear indication that the terms of the question are ill posed. In this case the concept of free will is misplaced. I will elaborate on this topic in the next chapters.

Why does the mother love her children? ...

Evolutionary psychology explains how a given species could have come to have a certain characteristic: through random mutation plus non-random natural selection. However, it does not tell us how this

3. Beyond Materialism

characteristic is produced in the single individual. How the activation of particular areas of the brain of a specific mother correlates with the feeling of maternal love.

After which it must be considered that neurosciences show us that there are specific areas of the brain, whose activation is quite correlated with maternal feeling. But correlation is not causation. Thus the nature of feeling remains entirely unexplained. We will go deeper into this topic in the next chapters.

Why are we sociable and some behave well? Where does morality come from?

The same observations made on the previous question apply here too. I only add that many of the problems of moral philosophy, that have emerged in the modern sphere, did not arise in the classical era (morals often based on the concept of virtue) or in Eastern philosophies (morals based on overcoming ignorance). For example, the moral problem of the "trolley" occurs only within the utilitarian type of morality. To call utilitarianism a morality is meaningless. In fact, what I am proposing here is that "the hard problem of consciousness" has an ethical aspect: one cannot derive a quality (in this case: a moral value) from a quantity. A mathematical calculation of costs and benefits cannot give rise to moral value.

Why is economic and social development? ...

I think that in recent years it became quite clear, to any person without specific conflicts of interest, that globalization based on liberalism has led to great planetary tragedies, which could have been perfectly avoided.

What is the truth ? ...

The theory of genes and memes is simply brilliant. Today, however, in the scientific environment it is considered partly outdated. It is, however, a good "thinking tool" for understanding and interpreting the world. I will not go into the specific reasons for criticizing these theories here. I think it is sufficient to recall here how in the book that introduced the concept of "meme": "The Selfish Gene", (a book much criticized by those who have never read it), the great evolutionary biologist Richard Dawkins explains how we are not slaves to our memes: because we can become aware of their presence, and take appropriate action to prevent them from spreading. It's what we do every day when we realize that we have a toxic thought in our mind and we resort to the suggestions of "folk psychology" to eliminate this thought.

As far as genes are concerned, it suffices to observe here that the very existence of genetic biology which verifies the possibility that a couple can have children with genetic problems, before giving birth to a child, is proof that we can try not to be slaves of genes.

How do you judge a behavior? …

Utilitarian ethics are just an aberration of ethics as imagined in classical times or in the East. I won't dwell on this topic here, because this is not the place, but for the interested reader I suggest to deepen the concepts of Dharma, gratitude and compassion (whose etymology derives from: to suffer with).

Why is there a universe like ours that has anthropic parameters?

The existence of the multiverse does not eliminate the problem, it merely shifts and complicates it. The multiverse tries to explain the fact that our universe has particular laws by saying that there is a universe for every possible configuration of laws. However, the multiverse itself obeys laws: for example, those that establish how its universes are born, "reproduced" and disappear. And who chose the laws of the multiverse? A multi-multiverse?

Is there any meaning in life? …

All the best philosophies of different cultures of all ages agree that the meaning of life is not to be understood as something found in the future. Suppose that a person dies young (or old) without having achieved certain vital goals such as starting a family, having a career in business, having written a book, having become rich and/or famous, then does his life have no meaning? The wise speak us of a way of seeing things, a way of perceiving and relating to the world here and now. Classical thinkers called it virtue, while Eastern thinkers called it enlightenment. Setting aside the time factor, any discourse or mechanical judgment (retroactive or not) on the meaning or lack of meaning of life

3. Beyond Materialism

automatically disappears. Because the meaning of life is not in the future, but it is now. A life that ends prematurely or fails to achieve certain goals makes no less sense than a long and successful life.

Why do I exist? …
A single person doesn't have much value evolutionarily speaking, that's obvious. But that's not the point here. Rather the question to ask is: Am I a separate entity from the totality of existence? Because if I'm not, then I exist for the same reason and value that all of reality exists. I will explore this theme further in the continuation of the book.

Who am I? …
The greatest philosophers and mystics of history agree in stating that: "Everything that can be perceived is not the one who perceives." Huang Po (Chan/Zen Buddhism Master).
We are nothing that can be perceived, we are that which perceives. My body is not the car I drive, but rather I am inside the car and I move around in it, but I am not the car. Iterating the argument, I am not my body, my body is to me as the car is to my body. In the same way, inside my car I find various objects (the steering wheel, the windows, the keys), but I am not them. In the same way I am not what I encounter in the experience of my body and my mind, I am not my thoughts: I observe my thoughts but I am not a thought. When I stop thinking about a mandarin, I don't stop existing. So I'm not the thought of a mandarin. Suppose we are looking at a panorama and then think "I am looking". Is the thought "I am looking" the one who looks? Similarly, is the thought "I feel" the one or that which feels? Behind any thought there is a single subject who thinks it, and this subject cannot be thought in turn. Because if it were thought it would not be the subject: a thought does not think, the thought "I know" is not the one or what known. But I do know, therefore I am nothing that can be known. Each of us is something more mysterious. We will be able to explore this topic further.

Yes, fine, but what is my essence? Am I not conscious?
Of all things, consciousness is the only one that cannot be an illusion. An illusion implies that the mental representation of an object does not coincide perfectly (or at all) with the object it is supposed to

represent. But consciousness is not a representation, rather: consciousness is a presentation. Consciousness (or awareness) begins to exist the moment it self-knows, the moment it is present to itself:

1. If I am aware then I exist, when I stop being aware then I stop existing.
2. If I exist then I am aware, if I cease to exist then I cease to be aware.

In consciousness, being conscious and existing coincide. Therefore consciousness knows itself not as it knows any other entity: objectively; "consciousness knows itself by being itself". Consciousness exists when it knows itself, and it knows itself when it exists. Being and awareness coincide.

How do we relate to the Amazon and the environment in general?…

Amazon, the planet earth, plants, animals, human beings, etc … do not have a mere utilitarian value. By delving into the theme of what reality is, we come to discover that we are not separate from the rest of the universe; we are not monads; we are manifestations of one thing. Hence the golden rule and universal compassion.

But are all these (materialistic) answers, albeit a little depressing, really correct?…

As already said: they are correct but incomplete. In this chapter I have tried to complete the answers given in the previous chapter, showing that in order to overcome them it is enough to see them within the larger context, from which they ultimately emerge.

When you don't understand something or when something seems unsatisfactory what you need to do is zoom out, and see the larger context in which they are immersed.

4.

CONSCIOUSNESS

The truth is simple. If it were complicated, everyone would understand it.

Walt Whitman

Of all the koans, "I" is the highest.

Ikkyu (Zen Master)

Consciousness is never seen, but is the Witness; consciousness is never heard, but is the Hearer; consciousness is never thought, but is the Thinker; consciousness is never known, but is the Knower. There is no other witness but consciousness, no other hearer but consciousness, no other thinker but consciousness, no other knower but consciousness. It is the Internal Ruler, your own immortal self. Everything else but consciousness is mortal.

Brihadaranyaka Upanishad (III.vii.23)

In the previous chapters we have developed the part destruens. The part construens of the book begins with this chapter. The compass we will use will be to see things in as broad a context as possible. While the method will not be "the doubt" as in Descartes, but the "sense of wonder" for the mystery as indicated in Aristotle's Metaphysics.

The starting point for overcoming mechanistic-materialism is provided to us by the father of mechanistic-materialism itself:

Democritus. Life is ironic, isn't it? Democritus must not have been an idiot at all, as he was often depicted in the paintings of European artists: where he often has the appearance of a drunk homeless man. As a matter of fact, if placed in the right context, the mechanistic-materialism of Democritus is the precursor of the atomist hypothesis in physics. The great physicist Richard Feynman was once asked what idea of physics he would like to preserve in the event of a total collapse of human civilization; so that the time to rebuild civilization would be as short as possible. Feynman replied that he would choose the atomic hypothesis of the world.

Well Democritus suggests the starting point for overcoming mechanistic materialism in the following dialogue (recreation of Galen's testimony collected in Diels-Kranz fr. 125):

The intellect says: "*The color is by convention, the sweet is by convention, the bitter is by convention... In truth there are only atoms and emptiness*".
The senses say: "*Miserable intellect... is it from us that you are taking the evidence for which you reject us? Your victory is your defeat!*"

Yes, well, brilliant, but what does it mean? It means that the conceptual entities with which we interpret reality, such as atoms and emptiness, are not known directly: they are not presentations but representations of objects that we assume exist outside our mind. These representations derive their evidence from sensible perceptions: colour, sweet, bitter ... The senses accuse the intellect of having first built a conceptual building on their foundations and then arrive at denying the existence of such foundations. The intellect builds a conceptual castle in which there is no longer any place for the sensible qualities on which it is built; sensuous qualities are considered a convention, names devoid of reality. The senses then warn the intellect that if they are eliminated, the whole building will collapse. If you eliminate the experience everything collapses. Because it is not the qualities that are names and conventions; but it is the concepts that are names and conventions.

4. Consciousness

But what is meant here by the term experience? It does not mean a memory of the past, or a particular skill acquired over time. Experience means everything we know directly: thoughts, emotions, sensations and perceptions. The flow of experience we call mind. While what the experience appears to, is defined as consciousness.

It is important here to underline the distinction that exists between mind and consciousness. The mind is the set of thoughts, perceptions, sensations, emotions, while consciousness is what the mind appears to. Each of us is her/his consciousness but not her/his mind. Consciousness is like the cinema screen on which the movie of the mind is projected.

Here Democritus seems prescient: hearing the intellect speak it seems to listen to the thesis of the modern emergentists, eliminativists and illusionists. The emergentists, eliminativists and illusionists affirm, with appropriate differences between them, that color, sweet and bitter (and all experiences in general) either do not exist at all or are illusions, a game of mirrors. Someone claims that since neurosciences have not been able to find the seat of consciousness in the brain then consciousness either does not exist or is an illusion. The impression of finding ourselves as spectators of the film of the mind inside the "Cartesian theater" of our brain is like an optical illusion.

Their argument can be summarized as:
1. With experience I discover that my body owns the brain. And this discovery can only be made through an experience: for example by putting on a helmet and performing an FMRI and visually seeing the resulting images on a computer screen.
2. Afterwards, examining the images on the computer screen, I realize that no matter how hard I search, I can't find my consciousness.
3. So I deduce that my consciousness, the one that is seeing the images obtained from the FMRI, does not exist.

Absurd isn't it? And here Democritus would let the consciousness says: foolish intellect who climbs a ladder resting on a cloud, and once at the top, removes the ladder.

The starting point of our philosophical "second navigation" (as metaphysics was called in the classical era) will therefore be a question about the nature of consciousness. In this book I will use the terms "consciousness" and "awareness" interchangeably. Some authors also use the term "presence". In this book I will use the three terms with the exact same meaning. If the reader is more comfortable, or more familiar, or has more insight, with the term consciousness, then feel free whenever you read the term awareness, to substitute it with the term consciousness. Of course, the reverse is also true. In the Buddhist tradition there is a tendency to use a somewhat different terminology. For example, Huang Po uses the term hsing which in Chinese means simultaneously mind, heart, spirit, soul, reality. In Western translations hsing is translated as Mind. But reading by the texts, it is clear that in the end Huang Po is referring to what here we call consciousness, and not what here we call mind. The latter is the set of thoughts, sensations, etc ... and is a manifestation in the Mind/consciousness. In the end it is enough to understand the meaning of the words.

In Plato's cave myth, it is said that people are chained inside a cave and are forced to look at images projected on the inner wall. It turns out that these images are reflections of objects passing in front of the entrance of the cave, on the side opposite to the wall. The images on the wall are pale reflections of the real objects outside the cave. But the prisoners, having known only the images, mistake them for the real objects. At some point, suddenly, a prisoner frees himself from the chains and turns to the opposite side and discovers the truth. The prisoner's turning around is the definition of the word "conversion", i.e. changing entirely (con-) the direction of observation (-vertere). The conversion proposed in this book is to go from seeing consciousness as a manifestation of the physical world to seeing the physical world as manifestations of consciousness.

But how does a conversion come about? In as many different ways as there are people; but almost always it is first aroused by some particular experience. Probably almost all of us have happened, in some given moment of life, to stop and notice the extraordinary nature of that phenomenon we call consciousness. In such moments we are struck by

4. Consciousness

the observation that consciousness is nothing and cannot be anything ordinary, but that it is precisely extraordinary. But, after we digress a bit about that observation, we get absorbed back into the flow of life. These observations can be aroused by the most varied experiences, for example by: life dramas, particular experiences during meditation, psychedelic experiences, lucid dreams, experiences out of the ordinary (ESP, OBE, NDE, etc...) or simple intellectual reflection. In my case, one of the most intriguing phenomena that has always prompted me to reflect is that of lucid dreams in episodes. It has always happened to me, and not infrequently, to have lucid dreams in which night after night the dream picks up exactly where it ended the night before. Just like watching a television series: where I realize it's a dream, which is quite vivid, where I'm often the protagonist, and where I can often choose the direction of the plot. With these types of dreams the difference between the waking state and the dream state is more blurred, and inevitably they have led me to wonder about the nature of reality and consciousness.

What is Consciousness? As a working definition we can say that consciousness is that to which experience appears. But if we ask ourselves to whom does the experience appear, we feel like answering with the word "I": consciousness is the I. And what nature does the I have? The answer to this question can be reached in two steps: the negative path and the positive path. First we find out what we are not and then what we are.

In Sanskrit the negative path is called "neti-neti" which could be translated as "not this and not that". It works by denying or excluding one by one everything that is not consciousness. We ask ourselves if the apple that we have before us on the table is consciousness: that is, if we are that apple. Obviously we are not that apple. We look around and discover that "I" am nothing that can I see, hear, taste, smell, touch. I am nothing that is external to my skin. So, instead, am I my body? Not even, because if I lose a limb, I'm always the same. Let's think about when the dentist gives us a partial anesthesia: the anesthetized part for a couple of hours doesn't seem to be part of our body. We touch the sleeping cheek and it feels like a foreign body like the apple on the table. A similar thing happens when we sleep on an arm, and then upon waking up the arm

37

remains asleep for a while: the impression we have is that the arm is not part of our body. If we were deprived of bodily sensations, the body would seem completely foreign to us. Evidently we are not our body.

So I wonder if I'm something that, somehow, is inside my body, maybe my brain or the mind produced by the brain. Let's try an experiment. If we ask ourselves: am I below or above the height of the stomach? We will have the clear sensation of being above the stomach. Are we located between the neck and the stomach or above the neck? Definitely above the neck. Are we between neck and ear height or above? Definitely above. Where do we seem to be exactly? Many of us would "feel" that we were somewhere between the eyes and an inch or two inside the skull. At this precise point, we feel our mind concentrated: the tangle of perceptions, sensations, thoughts and emotions that we call the "self", or "ego". This "self" is what is called "ahankara" in Sanskrit, and it is our little inner voice, the narrating voice in our thoughts, the one that says "I", which judges, which identifies itself with a thought, an emotion, an image of the self.

But if we "feel" this point, this ahankara, this little internal voice, if we are aware of this self, then we are not this self: we are not our mind, we are not the internal little voice. I am not this self for the same reason that I am not the apple in front of my eyes. Because I manage to separate it from the real me: to objectify it, to ideally place it in front of me and observe it. I am that which this self appears to. Then we realize that we have to distinguish the self (lower case), which is the mind, from the Self (upper case), which is the consciousness. The mind has an apparent physical location (between the eyes just below the forehead), but consciousness has not.

When we take any object and progressively exclude (concretely or ideally) every content, every part, which is not identified with the object, in the end there is nothing left, we arrive at nothing.

When, on the other hand, when we take consciousness and progressively exclude everything that is not the Self, we discover that in the end we do not arrive at nothingness, but that instead the "presence of an ultimate subject" remains. This presence of an ultimate subject, as devoid of content, we can call it as: the void, the silence, the unmanifest.

Buddhism especially develops this negative path in the form of the doctrine of the annata: the theory of the non-existence of the self or ego. This is an idea that was also discovered in the West by David Hume. If I try to find where the self is and meet the it, I will discover that I cannot find and meet it. All I can find and meet are just a tangle of emotions, thoughts, sensations that make up my personality. But I won't find any presence, any person. Has anyone offended me? Well, I don't find the offended person in my inner experience, I only find the emotion of being offended, without a subject. Am I happy? Yes, but who is happy? And where is she/he who is happy? Show me! I can't find the subject, the only thing present is the feeling of being happy.

Perfect there is no ego, no self, because I can't find it. Furthermore, I, the consciousness, the Self, am nothing that appears to me, I am no object that appears to the consciousness. Even concepts, abstract ideas and what they refer to, are objects that appear to consciousness. Therefore I also cannot be what a concept refers to. And now? Now begins the positive path, developed in an exemplary way in the traditions of Advaita Vedanta, of Kashmir Schaivism, and in some schools of Mahayana Buddhism.

The positive path consists in ascertaining that the objects that appear to my consciousness are not separate from my consciousness: they appear "in" and are "made of" consciousness. Schematically:
1. Negative path: consciousness is separated from objects;
2. Positive path: objects are not separate from consciousness.

There is a clear asymmetry between consciousness and its objects. Consider an object in front of your eyes, such as this book, and pass through the following steps:
1. The book consists of a shape with colors. Shape, by the way, is reducible to colors, because we infer shape from a difference in colors.
2. Color is not distinct (or separate) from "seeing" colors. Seeing means perceiving colours, and a color is nothing other than being

perceived. I cannot place myself ideally and/or concretely between color and seeing colour, because they are two different names for the same phenomenon in consciousness.
3. Seeing is not separate from being aware of it. I can't put myself between consciousness and "seeing", in order to see them come into contact with each other. Seeing is an aspect of being conscious. If there were no consciousness there would be no seeing: I see something as long as I am aware of it.

The same goes for other perceptions, sensations, thoughts and emotions. The only thing present in them is being aware of them: they are not separable from consciousness: ergo they are made of consciousness.

Realizing experientially what the positive path proposes is what is called the non-dual experience. That is, the realization that objects are not separate from consciousness, that they are manifestations of consciousness in consciousness. But in that case it follows that there is no physical world outside consciousness, for if it did exist then the objects of consciousness could not be "in" consciousness and "made of" consciousness, but only appear "to" consciousness. In contrast to how a jpeg file of an image appears "to" the computer but not "in" the computer (is there an inside and an outside of the computer that is not an arbitrary convention?) and is not "made of" computers.

The negative path is exemplified by the meditative tradition of mindfulness in which we observe the contents of our consciousness and place ourselves in the attitude of observers separated from these objects. Mindfulness is a dual meditation.

The positive path is exemplified by the meditative tradition of nididhyasana in which we observe the contents of consciousness as not separate from consciousness: as a manifestation of it. Nididhyasana is a non-dual meditation.

When I look at a rose, I also automatically have the sensation of being in front of that rose and of looking at it. This sensation is the ego, which creates an apparent subject as opposed to an object. But if by

4. Consciousness

meditating we manage to make this feeling of self, and every narrative voice, disappear for a moment, then only the rose remains. This time, however, not as an object opposed to a subject. The rose no longer has a collocation with respect to an ego subject. The rose is simply present, it is a simple manifestation of awareness, without any other objective or subjective element.

The idea of non-duality is both a philosophy and a form of spirituality; and it has been discovered more or less explicitly over the centuries by the most diverse traditions. For this reason Leibniz and Aldous Huxley have called it the "perennial philosophy".

One tradition in which non-dual understanding has been particularly elaborated is Advaita Vedanta. I believe that this is the philosophical-spiritual tradition that has by far come closest to the pinnacle of non-dual understanding. Much of the conceptual approach elaborated in this book is based on Advaita Vedanta. Advaita in Sanskrit means "not two". Why saying "not two" instead of just saying "one"? Because in their humility, the ancient rishis (seers) of the Indus valley, who discovered this understanding, understood that the intellect could not grasp this idea in a positive way and thus affirm that the ultimate reality is one, but only in a negative way and thus claim that ultimate reality is non-two. Saying not-two, best captures the asymmetry of the separation between consciousness and its contents.

All the different non-dual traditions decline their understanding in the language and culture in which they developed. Therefore in the way of explaining, there are different sensitivities and concerns. Also, in this book, non-duality is explained with a particular language and sensitivity, which are those of the culture from which I, as an author, come, and in particular from my scientific mindset. Therefore it is likely that it will resonate better with like-minded people as I am.

But the basic intuition of non-duality is always the same in the different traditions that have discovered it. An initial work of translating the languages is therefore necessary, from that of the tradition in which one learns to one's own personal one. Once the translation has been

41

made and the basic intuition has been learned, it must be understood experientially.

The intellect can come to apprehend this idea but never fully capture it. To capture it you have to experience it. For this reason, when the Buddha was asked about the nature of reality, he tended to smile and keep silent. His proposal was: walk the eightfold path and you will discover it for yourself. Lao Tzu expressed this same attitude when he said: "He who speaks does not know. He who know does not speak". Non-dualism is not a matter of belief but of understanding: there is no need to believe that reality is non-dual, because it can be understood, it can be seen, it can be directly experienced.

In the next chapters I will try to give some satisfaction to the intellect by presenting 10 arguments, or ways, that I have elaborated in favor of non-dualism. Reading the literature on non-dualism, to which I had access, I found that I missed reading well-structured arguments that fully satisfied my analytic mindset. The literature on non-dualism has as its first and last aim that of arousing the non-dual experience, beyond logical demonstrations. Direct experience is undoubtedly the most important thing and the ultimate judge of the value of this thesis. Except that due to my mindset, my intellect didn't authorize me to go ahead before having convinced him with its methods. And so I tried my hand at developing a whole series of arguments that would convince it. These arguments and the ideas expressed in this book have been incubating in my mind for several years, but curiously they emerged, took shape and were written within the last 15 days of January 2023; in those days I had the feeling that the book was writing itself, and I still have that feeling. The next 10 chapters are dedicated to the reader who identifies with me in this respect.

All these arguments have two aspects:
1. Each independently demonstrates that reality is a manifestation in consciousness.
2. Each argument responds at the same time to a set of possible objections, which may arise in it or in other arguments.

4. Consciousness

Due to the second aspect, the arguments complement each other. For this reason it is important that the reader read them all carefully.

Often an argument may be fully convincing, yet there may be some inertia left to embrace it completely. This happens when we realize, explicitly or implicitly, that the argument implies a paradigm shift affecting other aspects of our world view. Consequently we need to know how the new paradigm positions itself in relation to these other aspects. This results in a whole series of questions, which can take the form of possible objections. The arguments that I will illustrate and the following in-depth chapters try to cover all possible questions, or at least all those that have come to my mind. If I haven't been able to answer everything, I still hope that I have been able to provide enough conceptual tools for the reader to be able to construct the answers themselves. In addition, I have included at the end of the book a discreet bibliography of texts that have the ability, in my opinion, to make visible all the implications of the paradigm shift proposed in this book, so that the reader can handle them appropriately.

Perhaps in the end all of these arguments are expressions of a single underlying argument explored from different angles. Like the different paths or ways to climb the top of a mountain. In particular, the various arguments can be classified into two types according to whether they exploit the way of the negative path or the way of the positive path. After reading the arguments, the reader will realize which of the two paths each argument belongs to.

What must always be remembered is that non-dualism is not in conflict with science and everything we know about our universe. Science and non-dualism are placed on two distinct planes of analysis of reality. We will have the opportunity to elaborate on this theme in the next chapters. All this arguing does not serve to discover a corner of our mind next to others. Instead it serves to go beyond the mind. The concepts and arguments developed in this book serve to go beyond all other concepts and arguments, so as to stay with the single thing that remains: awareness.

5.

ARGUMENT FROM QUALITY

Concepts that have proved useful for ordering things easily acquire such an authority over us that we forget their earthly origin and accept them as immutable data.

Albert Einstein

The sensation of color cannot be explained by the physical's objective image of light waves.

Erwin Schrödinger

As a first approximation we can distinguish two types of objects of experience: concepts and qualia.

The concepts are generally presented to us in the form of verbal language and "represent", in the sense of "referring to", something else. The concept of "snow" does not exist in itself, what exists is the real physical snow outside my mind. The word "snow" makes proper reference to this entity external to my mind. In Indian philosophies they are called with the term "nama", which can more or less be translated as "name".

Qualia, on the other hand, are characterized by "representing themselves", i.e. they are "presentations": because they do not refer to anything other than themselves. John Locke divided them into primary and secondary qualities. Among the former we include mathematical entities, while among the latter we include perceptions, sensations,

5. Argument from Quality

emotions. In Indian philosophies they are all called with the term "rupa", which can more or less be translated as "form".

All we know is therefore "nama-rupa": names and forms. Concepts are always defined in relation to something else, they are symbols: in a dictionary each word/concept is defined in terms of other words/concepts, weaving and knotting a network where each node is connected to a set of other nodes that define it. Every concept is constituted or definable or reducible in terms of other concepts and/or other qualia.

Take light as an example: whatever physical light is, it is not what we perceive as light. When a stream of photons hits the retina, it doesn't pierce our skull and illuminate its interior: instead, it becomes a train of electrical impulses related to, but different from, physical light. The inside of our skull is always dark when we experience light. So photons are concepts that we've come up with to model certain kinds of perceptual regularities, and they work beautifully, but no one has actually directly perceived one of them. Strangely (from a materialistic perspective, that is), we perceive light in our night dreams, even if our eyes are closed and therefore the correlation between physical light and qualitative light is lost, and we understand thus that they are two very different things.

The very solid-looking physical table on which I am writing this book is just a name, a concept. In fact, physics teaches us that the table is made up of a tangle of atoms, which are in turn made up of subatomic particles, which in turn can be defined as probabilistic quantum fields. These fields are anything but mentally viewable: they don't have the visual appearance of my perceived table at all. So it is true that the physical table outside my mind is just an abstraction, a concept, a name. Totally different, then, from the visual image present in my mind: the perceived (or mental) table. Physics teaches us that by properly studying the table, we discover that it does not exist: what exists is a field of probability.

But I see the table in all its solidity! The table I see is within my awareness; it's not the physical one out there. There's just a probability field out there. Perfect, but in saying this, then, isn't there a problem similar to that of consciousness which, looking at an FMRI and not finding itself, says that itself doesn't exist? Let's analyze the logical sequence:

1. The image of a table appears in my consciousness: this is the mental table. We do not directly perceive the physical table.

2. Starting from the image of the mental table, common sense leads us to hypothesize that there is a physical entity, a physical table, independent of my consciousness which is identical and which somehow causes the image of the mental table.

3. If to this, we add appropriate hypotheses about the physical existence of particular measuring instruments and various considerations, we arrive at the discovery that the physical table does not exist, but what exists is another more complex entity to understand: a quantum probability field, to which we have no corresponding mental image, unlike the table.

4. But then, we are in the presence of an argument from absurdity. To prove the falsehood of a hypothesis X, what we do, is to assume that X is true, and then show that this assumption leads to an absurd (contradictory) result. If starting by assuming X, then we discover that this implies that Y exists, and that the latter in turn implies that X does not exist; then it means that neither Y nor X exist.

5. Therefore, if starting from the hypothesis that a physical table exists we arrive at denying its existence, then the hypothesis of the existence of the physical table and of the fields of probability out of my consciousness is at least "illusory". In the argument from substantiality, I will better clarify what I mean when I use the term illusion.

6. Probability fields exist outside my mind, but not outside my consciousness. I will elaborate on this topic in the next chapter.

If we can say that the physical table does not exist in itself, what can we say about the mental table? In general, what can be said about the consistency of mental objects, qualia? Here, it is good to stop for a

5. Argument from Quality

moment to clarify a key point of this argument. When we see the color red immediately we tend to think of it as a representation of a physical property, of an object external to the mind. That is, we usually, if not always, do not realize that qualia are presentations, not names or representative symbols like concepts. In this sense, it becomes legitimate to ask the question: what is a qualia in itself? What is it made of? What nature does it have?

We have established that red is related to, but it is not a particular spectrum of physical light. But then what is red? What is it made of? When I have a table in front of me and I wonder what it's made of, I can answer: it's made of wood. But how do I answer the question: what is a qualia made of?

Close your eyes and hear the noises around you. Wait a moment and then ask yourself: what are they made of? It is not worth answering that they are acoustic waves, because we are not talking about the physical model, just as we are not talking about electromagnetic waves. We are talking about what sound, as a quality, is in itself, and in general what any other quality is in itself.

The nature of qualia is different from that of concepts and therefore of physical entities. The word "existence" comes from the Latin ex- (outside) + -sistere (to stand): an entity exists when it distinguishes, it separates, itself on the background of reality, of being.

If an entity A has no appearance that distinguishes it from an entity B, then A and B are the same entity. Instead an entity C will start to exist and will continue to exist as long as someone can distinguish, can separate, it from all other entities. Asking for an explanation for the existence of an entity X means asking what allows it to distinguish itself from the rest of the other entities and thanks to what this distinction has emerged.

A disassembled car does not yet distinguish itself as a car: therefore it does not yet exist as car. Once one assembles it then it stands out as a car and will exist. All entities that allow an entity A to distinguish itself, are the causes of A. Who assembles the car (efficient cause) and the components of the car (material cause) are the causes of the car. An automobile is defined and explicable in terms of (it is distinguishable thanks to) its components and the factory that assembled it.

In order to exist concepts rest upon, i.e. they are explained in terms of (i.e. they are distinguishable thanks to) other concepts and/or qualia. On the contrary, the qualia are distinguished, i.e. they explain their existence, by themselves, a property that scholastic philosophy called perseity: the qualia have their "being per se". A qualia is therefore not like a car (which is a concept for the same reason that the table on which I write is, as we saw earlier) that has to be explained with other entities (components and factory).

One who has never seen red cannot know it by seeing green or blue (that is, other qualia) or by knowing the frequency of its related physical electromagnetic wave (a concept). While concepts can be fully understood based on other concepts (such as a car based on its components), this is not the case with qualia: the only way to truly know the taste of a mango is by eating a mango.

Qualia sit directly on the ground of being, while concepts are built upon this surface. However, conceptual networks cannot be entirely suspended in mid-air: at least some concepts must act as anchors and have a one-to-one association with direct and immediate experiences, the qualia. A language, to be understandable, must have a minimal percentage of its vocabulary that are nouns used to refer to qualia.

Given that a concept represents, that is: it refers to, something other than itself, then the categories of true and false can be applied to it. A concept is true or false to the extent that it correctly represents what it refers to. Instead a qualia being a presentation, meaning we could say that it represents itself, is beyond the categories of true and false: a qualia simply is, simply it is itself.

Now a materialist might say: okay, we could conceive of physical light as a concept, but perhaps this is just our cognitive defect and

5. Argument from Quality

actually physical light is a non-conceptual entity. To this, though, one can quickly answer: sure, why not; however, if physical light is non-conceptual, then it can be nothing but a qualia. There is no escape: either something stands on its own (and we call it a qualia) or it leans on others different from itself (and we call it a concept): tertium non datur. In the argument from concepts we will go into more detail about the nature/reality of concepts and show that they have a double aspect of name and form. We shall see that names and forms are in themselves qualia.

If the qualia have their own sufficient reason (they are distinguished from other entities by themselves), they are their own explanation, that is, they rest directly on the ground of being. Now, since consciousness is that to which qualia appear, it follows that consciousness (note: not mind) is the ground of being. Consciousness therefore does not have to seek an explanation for its existence on concepts (see: physical entities), but rather it explains itself. I'll develop what this actually means later.

Many of the logical, ontological and ethical paradoxes we have encountered are rooted in our willingness to imagine a language (that is, a conceptual system) without foundation, floating and unmoored, disconnected from the foundation of qualities.

From this kind of reflection originates the empiricism of the experimental method in science: a scientific theory must have a connection (direct or indirect) with something we perceive directly. A theory totally suspended in the void remains only an intellectual exercise.

In this light, the Hard Problem of Consciousness can be seen as a consequence of our clumsy attempts to reverse the ontological order, trying to ground qualities on concepts.

In the ethical field, the inversion of the ontological order translates into trying to derive moral qualities from a conceptual system. This inevitably generates a whole series of moral paradoxes. David Hume had already arrived at this conclusion when he stated that morality is a "matter of fact and not of abstract science". Trivially: who has ever tried and found inspiration in a moral rationalist? By moral rationalist I mean someone who bases his morality on an abstract conceptual system and not on moral qualities (values), such as love, compassion, kindness. The Italian word "saggio" as the English word "sapient", derive from the Latin "sapere", which means "to have flavour". A wise person is recognized because she has flavor, that is, she has human qualities beyond simple intellectual notionism. When you meet her, she not only transmits rationality but has a human thickness. Conversely, people who have tried to lead their lives with only rational morals have inevitably ended up having a utilitarian, arithmetical, accounting approach to life, which perhaps has allowed them to climb socially and economically. Except that once they achieved success they discovered they weren't happy, and then either they ended up saying that nothing makes sense (because success doesn't give meaning to life), or maybe they realize they have taken the wrong approach.

The inversion of the ontological order can have dramatic consequences in our level of spontaneity and light-heartedness. Should we base our lives on abstract ideas, or should our "lives" dictate those abstract ideas? Let's take the example of love. Should we base our ideas about love on our lived experience, or should our lived experience of love be based on our notions of love?

Clearly we must base our ideas on our lived experience rather than the other way around. Living the other way can cause a lot of problems in our lives, making us become preoccupied, rigid and immutable, preventing us from experiencing the spontaneity and richness of life.

The lived view is primary, ideas about it are secondary. Warning: ideas are secondary, but not useless.

Even the problem of evil finds a new light if framed in the perspective of the inversion of the ontological order. There is practically no philosopher or religion that hasn't come up against the problem of evil, and hasn't tried to find a solution, a theodicy. In monotheistic religions it is a particularly felt problem: why does evil exist if God is good and just? Most solution hypotheses are a variant of the following

5. Argument from Quality

basic pattern: evil serves to obtain greater good in the future. This basic pattern is then dressed differently by different cultures and traditions with a castle of concepts and explanations. However, the basic problem remains and can be summarized as: why do the qualia of pain and suffering exist?

We have established that qualia have their being per se, and therefore cannot be explained with concepts. To explain them with concepts would once again be an attempt to invert the ontological order.

Qualia such as pain and suffering cannot be explained, but can instead be transcended. Evil is not explained, it is transcended.

A qualia, A, is transcended when one experiences its dissolution and transfiguration into another qualia, B, of which A is an expression. Only by having the experience of dissolution-transfiguration do we perceive the meaning of pain. Just as the taste of mango cannot be communicated in words and explained conceptually, the dissolution-transfiguration of a pain-suffering cannot be explained, but only experienced in the first person. The transfiguration of pain is a consequence of the fact that the qualia add up to each other in an intensive rather than an extensive way. An example of intensive summation is that of musical notes to make a symphony. A note taken by itself may be unsightly, but when properly inserted into a symphony it acquires a beauty it would not have had on its own. The unpleasant note can be vaguely recognized in the symphony, but now it has a sublimated character, which I could never have perceived in the single note. There is an episode of Tolkien's Silmarillion fiction, in which this phenomenon is described very well. In the Silmarillion there is an evil deity named Morgoth who together with a number of other deities, the Ainur, must compose a symphony. The symphony is the metaphor of creation. The deities take turns adding their beautiful notes and all goes well, until it is Morgoth's turn to add his own note. However, its note is out of tune, making the so far composed symphony rather unpleasant. The introduction of Morgoth's note is the metaphor of the entry of evil into the world. After which comes Eru Iluvatar, the supreme deity, who welcomes the symphony of the lower deities and sublimates it with suitable adaptations. The end result is a work of extraordinary beauty that manages to transcend and transfigure the note of Morgoth. But there's more: the final result would not have been as sublime without this note. This Tolkien myth explains to us that evil cannot be explained, but it can be transfigured, and that only if we live the experience of transfiguration can we understand it. But before this experience, evil is totally incomprehensible. I will return to this theme briefly in the chapter on the experience of meaning.

6.

ARGUMENT FROM SIMULATION

Theories have four stages of acceptance:
1) this is worthless nonsense;
2) this is an interesting, but perverse point of view ;
3) this is true, but completely irrelevant;
4) I've always said that.

JBS Haldane

The authentic value of a human being is determined above all by the extent and by the way in which he freed himself from the self.

Albert Einstein

In the previous chapter we clarified several aspects of the nature of experience. Through an analysis of consciousness and its contents we understood that consciousness is not an illusion and cannot be explained with physical concepts and entities. On the contrary, consciousness has its own explanation in itself.

The concepts and the physical entities that some concepts represent are instead quite reduced in their ontological status. However, it remains to be clarified what this ontological status of theirs then is. In this chapter I will introduce a new argument that will clarify this aspect.

This argument finds its inspiration from the world of fiction, where entire conceptual universes (with their own internal structures, natural laws and inhabitants) are created in the form of novels, videogames, films or role-playing games. In some cases we even create nested universes, as in "The NeverEnding Story" by Michael Ende or in the magnificent film "The Thirteenth Floor", where imaginary or virtual realities are created or simulated within other imaginary or virtual realities, as in a system of Russian dolls, in which inside one doll one can always find another.

So let's try to analyze the architecture of these conceptual worlds. For brevity and ease of language we refer to a conceptual world as a simulation. Given the current pervasiveness of novels and video games, this linguistic choice allows us to "visualize" more easily what we are talking about and thus have a better intuition of its aspects.

Let us observe first of all that in every simulation there are 3 realities at play:
A. The simulated reality itself;
B. The reality in which the simulation is performed;
C. The underlying reality (which supports) the previous two.

So happens that reality C always coincides with reality B, and can never coincide with reality A. This means that the entities inside B and C can interact directly with each other, but the entities living in A cannot interact directly with those of B. Let us say that the reality-A does not have its being-in-itself, that is: it does not exist in itself; rather it exists in(side) other (actually-BC). Reality-A is an extrinsic appearance of reality-B. The property of existence in itself is called "inseity" in scholastic philosophy. Entities having this property are called "substances". Furthermore, reality A is not separable from reality-BC; on the contrary, the reality-BC is separable from reality-A. There is an asymmetry in the separation between A and BC. By separation I mean the ability to continue to exist independently: if BC ceases to exist then also A ceases to exist, but conversely if A ceases to exist then BC must not cease to exist.

Let's try to give concrete examples to understand this discourse. If we apply this architecture of the 3 realities in a landscape painting, we have:

6. Argument from Simulation

A. Reality-A: the mountains, valleys, rivers and sheep represented;
B. Reality-B: the canvas and the oils with which the picture is made;
C. Reality-C: the universe in which the painter lives.

Clearly the painter (reality-C) belongs to the same universe as the canvas and oil colors (reality-B). The painter is not a character in the picture (reality-A does not coincide with reality-B), and cannot interact with the sheep painted on the canvas.

The nature of the simulation is always highly dependent on the means available in the reality where the simulation takes place, and these can change over time: canvas and paint have been around for centuries, while computers are a recent development. If we take a video game as another example, we have:

A. Reality-A: the racing cars simulated in the video game;
B. Reality-B: the hardware on which the video game software runs;
C. Reality-C: the universe in which the game programmer lives, as well as the factories that produced the hardware.

The game programmer and microchip factories (reality-C) belong to the same universe as the hardware (reality-B), and clearly the programmer does not coincide with an avatar within the video game: if the avatar dies, the player does not die. Here we note the asymmetry in the separation between A and BC: if an avatar of the video game dies, the computer on which he is playing does not cease to exist. Conversely, if the computer breaks, then the avatar also ceases to exist.

Let us now take as an example that of a novel. In a book or movie, such as "The Lord of the Rings", we have:

A. Reality-A: Frodo "lives" within the film and is subject to (apparent?) causal relationships with other entities in the film (Sauron could kill him);
B. Reality-B: the television screen where the viewer is watching the film;
C. Reality-C: the universe in which the spectator who watches the film lives.

The spectator (reality-C) shares the same universe with the television screen (reality-B), but neither of them can be killed by Sauron, while Frodo can be (since both Sauron and Frodo are in reality-A). Frodo is not separate from Sauron and the television set; but the television set is separated from Frodo and Sauron.

Let us now consider a robot, which "perceives" the surrounding reality through a camera and reacts appropriately to these inputs. Later in the book I will explain how the perception in our consciousness is a radically different phenomenon from the perception that a computer can have.

A. Reality-A: The internal representation that the robot makes of the surrounding reality. For example a jpeg file for images and an mp3 file for audio. All the robot knows are these files, the robot doesn't know the external reality. The robot doesn't know the starry sky, the robot only knows bits of information: 0s and 1s. On the contrary, a human being knows the starry sky: it's black color and shining stars (these are qualities and not quantities like bits).

B. Reality-B: The robot's hardware, including its processor and memory unit, in which what the robot knows, reality-A, appears in the form of electrical switches.

C. Reality-C: The physical reality that surrounds the robot: for example the beach it is looking at or the starry sky.

The robot's reality-A is populated by objects that map some of the objects of reality-C (surrounding the robot) with a one-to-one relationship, but they don't coincide with them: they are just digital twins. The jpeg photo file inside my computer is very different from the actual photographed beach. Again, in this example the robot hardware (reality-B) belongs to the same reality that surrounds the robot (reality-C). While the reality-A, which appears on the reality-B, although related to the reality-C, is not of the same nature as the reality-C.

Let us now apply this model to conscious perception:

A. Reality-A: the physical world that I seem to perceive and in which my ego finds itself immersed, a world that I interpret as populated by entities such as mass, energy, space, time, fields, tables,

6. Argument from Simulation

chairs, planets, galaxies, etc… and completely describable in "quantitative" terms.

B. Reality-B: my consciousness, in which reality-A appears in the form of my perceptions, sensations, emotions. Ultimately a reality entirely populated by "qualitative" entities.

C. Reality C: the underlying reality.

It follows that the underlying reality (reality-C) and my consciousness (reality-B) belong to the same type of reality, populated by qualitative entities: the qualia.

If we could peek outside of our own consciousness, we would not find physical entities: we will only ever find consciousness itself, which then turns out to be the reality within which the "simulation of physical entities" ultimately exists. Physical entities have no inseity, but have their being (their existence) in consciousness. Consciousness is the substance of physical entities. In this sense, physical entities are manifestations of consciousness.

Let us pay attention and remember one fundamental subtlety. Consciousness is not mind: mind appears in consciousness like physical entities do.

Let us stop for a moment to understand the connection between the argument from qualia and the argument from simulation. The qualia argument tells us that physical entities, as concepts, need to be grounded in a qualitative substratum, otherwise they remain suspended in the air. The simulation argument tells us that within physical reality, physical laws can perfectly explain various phenomena without having to resort to assuming the existence of consciousness.

Physical entities form a coherent conceptual structure that lives in a type-A reality, and within that reality they do not need to postulate consciousness to be explained. Physical entities are like those that populate the universe of a well-constructed video game, in which the rules of the game form a closed referential system: they need not refer to anything outside their universe. This is why it is problematic to study consciousness (first person perspective, reality-B) from a scientific reality-A (third person perspective).

However, physical entities (reality-A) cannot be grounded on themselves. Since the reality underlying them is qualitative in nature, therefore consciousness has ontological primacy.

Following the simulation argument, one might ask: could the reality of consciousness (reality-C) itself be a simulation (reality-A) within an even more fundamental reality? In technical terms we ask: is consciousness a substance?

The argument from quality shows how grafted realities can only be populated with "concepts" ultimately grounded in qualia, while qualia are, by definition, grounded in themselves. Metaphysically we can say that: since the contents of consciousness (the qualia) have no perseitas, consciousness has inseitas. Therefore, consciousness is a substance and has nothing more fundamental to refer to.

Another intriguing line of thought focuses on what happens "at the edges" of a type-A reality. In a well-designed A-reality, you shouldn't notice any strange behavior, anything that might violate its internal rules; not a single clue that unmasks the underlying B-reality. A suggestive hypothesis to explain anomalous behaviors of our physical reality, such as quantum mechanics or paranormal phenomena, is that these occur near the "jagged limits of the simulation".

7.

ARGUMENT FROM REPRESENTATION

We are like spiders.
We intertwine our life and then move in it.
We are like the dreamer who dreams and then lives in the dream.
This is true for the entire universe.

The Upanishads

Life isn't about achieving a particular goal, it's about who we become in our path. Life is not about arriving but enjoying the journey. Don't be afraid of life. Believe that life is worth living and your belief will help create the fact.

William James

What has been established up to now may be quite unorthodox when compared with the idea commonly taken for granted today about how the perceptive process takes place.

In this chapter I will therefore analyze the perceptive process and, while we are at it, I will introduce a further argument in favor of the primacy of consciousness over matter. I will show that, here too, things must be seen in a broader context.

If we were to schematize the mechanism of the perceptive process, it is very likely that we would think of what in technical terms is called the "representational theory of perception" (RTP). The assumptions of the RTP are the following:

1. There are physical objects, and they exist regardless of whether they are observed by a consciousness. This is, of course leaving aside certain interpretations of quantum mechanics.
2. Physical objects possess a whole range of properties.
3. Some of these properties are observable by our sensory organs.
4. The sensory organs communicate the collected information to the brain.
5. The brain, in a way somehow still not fully understood, translates the information of the senses into experiences: that is, qualities such as colors, shapes, fragrances, etc. These experiences appear in our consciousness...
6. Our consciousness never observes the observable properties of physical objects as they are in themselves in physical objects, much less as they appear to the senses or as they appear to the brain. Our consciousness knows only their translation into qualia.
7. However this translation preserves a certain similarity between qualia and the corresponding physical property. This similarity is the factor that produces the "representativeness" of the representational theory of perception.

The RTP, thus described, seems very plausible. Furthermore, in the light of certain modern technological advances it seems decidedly indisputable. For example, audio and video recording and communication technologies give every impression that they can only be explained in light of the correctness of the RTP.

However the RTP in point 5 has its weak link: the translation of physical properties into qualia. There is as yet (and now we shall see that there may never be) a theory that explains this translation. This fact alone is sufficient to invalidate the RTP and simultaneously validate the non-dual model. But let's try to argue in more detail.

The conceptual scheme of RTP starts by assuming physical objects and goes back to consciousness (except for making a leap in point 5). Now let's try the reverse path: starting from consciousness to get to physical objects.

7. Argument from Representation

1. There is consciousness, and its existence cannot be an illusion. The existence of consciousness is the most certain fact a person can ever have. We can doubt the existence of physical objects external to the mind, but we cannot doubt the existence of consciousness.
2. Furthermore, there is no doubt that consciousness perceives qualia.
3. Consciousness can only know qualia: what is not a qualia cannot be known by consciousness. Here it is enough to do a little introspection to realize that everything that appears to our consciousness has a qualitative character. This statement mirrors the number 5 of the RTP.
4. Qualia can only exist in consciousness. In fact, no qualia is a physical property or a combination of physical properties. Color is not a physical quantity. The fundamental physical quantities are well known and are: length, time, mass, temperature, etc … None of these quantities or combinations is a colour. Just as it is not a sound, a fragrance or any other quality.
5. Qualia cannot be emergent phenomena from a physical substrate. Let's take the flock of starlings (swarm) as a typical emerging phenomenon. We note that its behavior is:

   A. Perfectly deducible from the behavior of its components. In fact we can simulate it on a computer. The single component does not behave like the swarm, but the behavior of the swarm can be deduced from the set of behaviors of the components.

   B. The behavior or property of the swarm can be described with the same physical dimensions that describe the behavior of the components. The fluidity of water is not present in the single molecules of water; and yet the primitive physical dimensions which describe water are the same as those which describe the single molecule of water.

Instead color (and any other qualia) CANNOT be an emergent property of brain circuits. In fact, a color cannot be described with the same physical dimensions that describe the neural network of the brain (space dimensions, frequency of the electromagnetic wave, etc..). Then:

A. Qualia are not physical entities and do not emerge from anything physical.
B. Qualia can only exist in consciousness
6. Therefore:
A. Since qualia are not physical entities and do not emerge from anything physical, qualia cannot be translated from a physical property and therefore cannot represent a physical property.
B. Since qualia can only exist in consciousness then there can be no flow of information between physical objects and consciousness. In fact, the exchange of a flow of information between an emitter and a receiver presupposes that these two share the physical dimensions of the flow. That is, the flow entity must exist both outside and inside consciousness. But within consciousness there are only qualia, which cannot be outside consciousness. Thus there can be no flow of information to and from consciousness. Therefore there can be no similarity between qualia and physical property and therefore a qualia cannot represent a physical property. Let's try to give a concrete experiential example. Let's imagine to be eating a mango and to savor its peculiar taste. This taste is like nothing found in the physical object mango. This happens analogously by comparing any qualia with any external physical object: there are no relationships of similarity. The "outside" has no trace of the mango flavor; maybe we would have chemical molecules specific to mango, but we will never identify the physical entity "mango flavor".
C. It also follows that what is outside consciousness is not available to be compared with what is in consciousness. No comparison can be made because it is, as explained in the previous point: there cannot be a flow of information between outside and inside consciousness. But if there can't be a flow of information between in and out of consciousness, then we can't say that external objects exist. To claim otherwise is just an imaginative exercise with no truth value, and violates Occam's principle of parsimony.

7. Argument from Representation

On the basis of this discourse it is clear that objects external to consciousness cannot be said to exist. On the contrary what we can say is that:
1. Physical objects exist "within" consciousness.
2. Physical objects exist "outside" our minds.
3. Consciousness cannot be explained in physical terms.

In the next chapter I will clarify what is meant by inside consciousness and outside the mind.

8.

ARGUMENT FROM ASYMMETRY

Anything known to you is separate from you.
Bhagavad Gita chap. 13

Gold is not an ornament, but an ornament is nothing but gold.
Sri Ramana Maharshi

If you can know it then it's not you: the conscious subject is separate from the known object.
If it is not you then you cannot know it: the known object is not separate from the conscious subject.
Alessandro Sanna

In this chapter I will develop an argument for non-dualism which is one of the most spiritually interesting to me. Once well and carefully understood, this argument becomes extremely intuitive and accessible experientially, not just intellectually; hence its spiritual value.

An interesting difference between matter and consciousness is divisibility, granularity. A material entity can be divided, concretely or ideally, while consciousness cannot be divided. Matter is therefore said to have parts while consciousness has no parts. Matter is extensive, while consciousness is intensive. Material objects begin and cease to exist when their constituent parts begin or cease to aggregate. Conversely, consciousness having no parts cannot begin or cease to exist, it simply is.

8. Argument from Asymmetry

When we question the existence of our body, we can do so by considering it one part at a time. I can ask myself if my hands exist, then I can ask myself if my head exists, etc… each part independently of the other, and each part perhaps further subdivided into other parts. Instead, when I ask myself if "I", my consciousness, exists, I can't do it in parts, it's all or nothing: either I exist or I don't exist; it's not a matter of degrees. Therefore "I", my consciousness, clearly appears to be a simple and non-compound entity. Matter has separations within itself, while consciousness has no separations within itself. Matter is granular, while consciousness is whole.

But what about multiple personality disorder or the split-brain phenomenon, where there would seem to be multiple minds at once? First, it is not clear whether these phenomena imply the presence of more than one mind or not. For example, in multiple personality disorder what is witnessed is that one has the sensation of a further "presence" next to one's own. Instead in the split-brain it must be considered that although the corpus callosum is severed, the two hemispheres can continue to communicate with each other through the nervous structures below the cerebral cortex. This allows to maintain a certain degree of integration of the brain and therefore, at least in part, also maintain a single mind. However even if the mind were to be able to subdivide, we are not discussing the mind here, but the consciousness. And from the point of view of consciousness when a person with multiple personality disorder, or split-brain is questioned about his sense of unity, she will respond as a normally gifted person: he finds no boundaries, no barriers, no sense of "this is in and this is out" in her experience. Her experience remains seamless, without interruptions.

The attentive reader may be wondering: how is it possible that consciousness does not have internal separations given that different contents appear to it: perceptions, sensations, thoughts and emotions? To explain the relationship between consciousness and its contents the Upanishads, the ancient sacred scriptures of India, as well as several other non-dualistic texts, use a number of metaphors. The most common metaphors are:

1. Consciousness is like the sea and its contents like the waves. What this analogy want to make clear is that the waves appear in the sea and are made of the sea. That the nature and substance of the waves is the sea, and that in the waves there is nothing but the sea.

2. Consciousness is like water, mind is like the sea, and content are like the waves. This is a variation of the previous one, which wants to underline the fact that consciousness is the material cause of its contents. Waves and the sea are made exclusively of water. So the contents of consciousness are made of consciousness.

3. Consciousness is like gold and its contents are like jewels. Or: consciousness is like clay and its contents are like various kinds of vessels. This analogy carries the same meaning as the previous one: jewels (or vases) can have any desired appearance, but their substance is unique: clay. Ramana Maharshi used to say: "Gold is not an ornament, but an ornament is nothing but gold".

4. Consciousness is like the mirror and its contents like the reflected images. Several non-dual modern authors compare consciousness to a cinema screen (or computer or television) and its contents to a film projected on the screen. What they want to convey with this analogy is that the film, or the reflected image, lives on the screen, or in the mirror. That the film has no texture, no reality, off the screen: see the simulation argument.

These analogies, like all analogies, can only go so far. And that point is the fact that you are comparing consciousness and its contents with 2 physical entities. So let's take a good look at where these analogies hold up and where they no longer hold. There are 3 aspects to consider: X appears "to" Z; X appears "in" Z; X is "made of" Z. Let's look at each aspect.

X appears "to" Z, or also: X is "on" Z.
Here the analogy is quite complete, and in fact in the following statements the sense in which the preposition "to" is used is the same:
1. the image appears to the mirror like
2. the contents of consciousness appear to consciousness.

X appears "in" Z, or also: X is "contained in" Z.

8. Argument from Asymmetry

Here the analogy ceases to be complete, and in fact in the following statements the sense in which the preposition "in" is used is different:
1. the image appears "in" the mirror unlike
2. the contents of consciousness appear "in" consciousness.

In fact the image is contained in the mirror in a different way from how the contents of consciousness are contained in consciousness. The image is contained "spatially" in the mirror. There is therefore a physical quantity, the length, which allows the image to be distinguished from the rest of the mirror. In fact, the image is close to the mirror surface. The light-reflecting atoms of the mirror are all a few nanometers below the ideal surface that bounds the mirror. Being contained, therefore, has a purely "extensive" meaning: that is, it can be characterized in a quantitative and dimensional sense by a physical quantity. On the contrary, the contents of consciousness are contained in consciousness not extensively but intensively. The image of a physical object that appears in my consciousness does not have a quantitative character with respect to consciousness. In fact, I am unable to say "where", within my consciousness, the image appears. I simply notice the presence of the image without knowing how to place it within a reference system. I can imagine a reference systems inside the image and therefore be able to say that the table is at a certain distance from the window; but I cannot place the table and the window in relation to consciousness. Same discourse can be made with a sound, a taste, a smell etc… Close your eyes and listen to the surrounding sounds and try to define a point with respect to your consciousness where they appear: you cannot succeed. The closest physical analogy to this condition is to imagine an ideal mirror with zero thickness. With such a mirror it would not make sense to ask how far away the image is from the mirror, because: since the thickness is zero, the concept of distance disappears. Unlike this, the objects appearing in the mirror have relative distances to each other; however they all have zero distance to the ideal mirror. Consciousness is not an "extensive container", but rather an "intensive container".

I call this dimensionless way of being of objects in consciousness **transparency**. The fact that known objects (that is, objects present in consciousness) do not have a reference system with respect to consciousness implies that they do not appear on (or to) the boundary

67

surface of consciousness with the external world. On the contrary, the objects appear totally immersed "in" consciousness. Similarly to how a physical object is completely immersed in physical space; apart from the dimensional aspect of the latter. Therefore, consciousness knows objects by observing its own internal conscious space, and not an external space: **consciousness is self-transparent**, it can observe, and actually only observes, its own interior. Conversely, a physical system such as an eye, a camera, a sensor, etc…, always and only observes its exterior. A camera records what is in front of its lens, not its internal circuitry. The camera lens is the outer boundary surface of the camera. It could be argued that the image on the retina is internal to the eye; however: the retina and its cells, as well as the image projected into them, have a dimensionality: they are each external to the other. We will return to the self-transparency of consciousness in the argument from ineffability.

X is "made of" Z.
While the image is made of photons, the mirror is made of atoms. Image and mirror are made of two different and distinguishable things. Technically we say that image and mirror are "separate" in the sense that we can find a separation between the two, and that each physically exists on its own. If you want to be physically precise when I look into a mirror I am not seeing the mirror, but the light reflected from the mirror. To get "in touch" with the mirror I can't use my sight, but I should use my touch and touch it. The image is not made of mirrors. And here comes a key and interesting point of the argument. When I look into a mirror and see an image I infer that there are other physical objects besides the mirror. Since light is not made of a mirror, then it has a different origin from the mirror and this allows me to say that there is a reality outside the mirror that allows us to explain the existence of light: that is, it is the object that is reflected in the mirror. Since image and mirror are separate things, then there must exist a third separate object that explains the image reflected in the mirror. If I see an image of an orange in a mirror, then there must be a real orange somewhere in front of the mirror.

If instead the image were made of a mirror, if mirror and image were the same thing, then I could not suppose the existence of a third

8. Argument from Asymmetry

entity separate from mirror and image. A physical analogy of this case would be that of a mirror in which a picture has been painted. The painting is inseparable from the mirror, and therefore I cannot assume that the painting is the reflection of a third object separate from the mirror and the painting. This analogy is only explanatory and should not be pushed further than necessary. In fact, if we want to be physically precise, even the painting on the mirror is separate from the mirror, because it is a varnish deposited on the mirror: the atoms of the varnish are distinct from those of the mirror. And so I have to imagine a whole host of other entities that produced the paint and made the painting.

In general terms: if in an entity A one observes aspects X separate from A, then there must exist an entity B, external to A, which explains X.

Let's take the case of waves: these are spatially contained in the sea and are mainly composed of water and air. The fact of finding air in addition to water allows me to infer the existence of a further third entity, the atmospheric wind, which explains the presence of air in the waves.

Let's take the case of a computer connected to a camera and equipped with artificial intelligence software that allows it to recognize images. What is the relationship between the computer and the images of external objects? The same that exists between mirror and image. The image is contained in the computer extensively and is not "composed of computers"; among other things, as we shall see in the argument from substantiality, if one searches for a substance called "computer", one discovers that it doesn't even exist. The image will be found in the computer in a jpeg type file format in a specific area of the RAM: there is a physical quantity (the memory address) which places the image in relation to the RAM. Furthermore, the jpeg file is not composed of the atoms of RAM: it is separable from the RAM. In fact we can copy the jpeg to some other memory medium. Now since the jpeg file is extensively contained in the RAM, it is not "constituted" by RAM, so its existence can only be explained by assuming other physical entities, distinct from the RAM: ultimately the physical objects filmed by the camera.

A different case is that of consciousness and its contents. For the latter are not separable from consciousness. Let's close our eyes and try to listen to the surrounding sounds. Can we establish where exactly sound ends and hearing begins? We fail: sound and hearing are the same phenomenon in consciousness. The known (sound) and the knower (hearing) coincide. While in the case of the mirror the known (the image) and the knower (the mirror) do not coincide. While a mirror encounters things separable from itself, photons from the reflected object, consciousness finds nothing it can separate from itself. Sound cannot be separated from hearing: they are two different names for the same phenomenon. Conversely, consciousness is separated from its contents: in fact, there are no contents that are always present in consciousness.

In this regard, the physicist Erwin Schrödinger made an emblematic observation: "the world is given to me only once, not one existing and one perceived. Subject and object are one". This observation tells us that reality presents itself to us only once and not twice. It does not present itself to us for the first time as a reality perceived in the subject: hearing; and a second time as an existing reality out there, the object: the sound. I never experience a subject and additionally experience an object. I have only one subject-object experience: hearing-sound. Subject and object are inseparable, they are the same thing: they are the same experience. I never first experience the subject, of hearing, and then the experience of the object, of sound. I have a single subject-object experience: hearing-sound. All I know of the object, sound, is in the subject, hearing, and vice versa. The distinction between subject and object is a concept formulated a posteriori, which does not come from direct experience.

Even the experience of the belief that there is a reality outside consciousness: that is, the experience that subject and object are distinct experiences, is only a further experience, which as such exists only inside my awareness. I never come into contact with an object outside

8. Argument from Asymmetry

awareness. The contrary belief is a further object always internal to awareness.

Since the contents of consciousness are present intensively in consciousness, i.e. consist exclusively of consciousness, then consciousness is all that is needed to explain them, they do not require an external entity to explain them. As was the case with the mirror image of the orange. There is therefore no physical world outside of consciousness. Everything is consciousness.

Consider a video surveillance system. A camera records what happens in a room and a computer analyzes the footage. In this example we clearly distinguish the known object, the knowing subject and the activity of knowing. The known object is the film. The computer is the knowing subject. Finally, recording the film is the activity of knowing. The cognitive process of the video surveillance system is a triad composed of three elements: knowing subject, known object and cognitive activity. These 3 elements are clearly distinct and separate from each other. More generally we can state that in any physical reality in which a cognitive process can be found, the cognitive triad is always present. The fact that there is this triad: that is, that object, subject and cognitive activity are 3 separate things, immediately implies that there is a reality existing outside the subject, a reality to which the subject has access.

Conversely, in your direct experience, when you know anything the triad collapses into a monad: subject, object and cognitive activity are a single indivisible substance. Check for yourself right now. Try, for example, hearing a sound. Realize that you are not at all able to separate between them: hearing (cognitive activity), sound (known object), and the awareness you have of the sound/hearing (the subject). At this point, and unlike what happens in a physical process, it appears that the subject-awareness in the act of knowing an object is not accessing a reality external to itself, because the object coincides with the subject, it is not separate. Therefore, by knowing an object, consciousness is knowing only itself, nothing distinct or external to itself. All reality is a manifestation of and in consciousness. Everything is consciousness.

Altogether we can state the following 4 fundamental principles of non-duality:
1. Consciousness is separate from its contents.
2. The contents of consciousness are not separate from consciousness.
3. Consciousness is not separate in itself, it is not granular. Therefore consciousness is a substance.
4. The contents of consciousness are separate in themselves and among themselves: they can be broken down into other contents, they are granular. Therefore the contents of consciousness are not a substance.

There is an asymmetry between consciousness and its contents: consciousness is separated from contents, while the latter are not separate from consciousness. If you can know something then it's not you: the conscious subject is separate from the known object. If something is not you then you can't know it: the known object is not separate from the conscious subject.

Contents (or objects) of consciousness include both internal mental states and physical phenomena external to the mind. Furthermore, the contents are separate from each other, so they are dual to each other. But not being separate from consciousness then they are non-dual to consciousness. This asymmetry is well exemplified by the word Advaita, which means non-duality, or non-two. In fact, reality is not multiple: it is not dual, but neither is it exactly one. Therefore one can try to capture this asymmetry by saying that "reality is not two".

At this point, given statements 1 and 2 or statements 3 and 4, it follows that: consciousness is not a physical phenomenon, and that everything is consciousness. Statement 1 constitute the negative path: the neti-neti: consciousness is not a thing, it is not an object, it is the void of which the Buddhist texts speak about. Statement 2 constitute the positive path: consciousness is the "whole" of which the Vedantic texts speak about. Taken together, the 4 statements mean: "emptiness is everything".

8. Argument from Asymmetry

Statement 4 embodies the empirical fact that any entity, but awareness, is granular. Therefore no entity has an "inside", an internal space, which can contain another entity. Every entity is outside, separate, distinguishable, from every other entity. Statement 3 denies granularity to consciousness. Consciousness has no separations in itself, it is a substance, it is a space that can contain, that can take on, infinite manifestations. The intensive space of consciousness is not to be understood in a physical sense, as an extensive space, as when I put water in a glass. Because water and glass manifest themselves separately: they are two distinct manifestations. Furthermore, everything that can be manifested about the water present in the glass, from the chemical composition, to the weight, to the transparency, to the electrical conductivity, to the viscosity, to the PH, to the single proton or quark that composes it, is a manifestation distinct from that of the glass and of the water itself. The "intensive space" of consciousness contains water, and the glass and any other thing X; in the sense that the "manifestation of X is the manifestation of the awareness I have of X". Any manifestation is always totally immersed, accompanied, contained, constituted, composed, realized, formed, inseparable from the manifestation of being aware of it.

Wanting to enclose this reasoning in a few passages, we can argue as follows. The gold (material cause) of a ring (effect) is separable from the ring, but the ring is not separable from the gold. In this way it is recognized that the gold is the material cause of the ring and not the other way around. Indeed, the material cause is separable from its effect, but the effect is not separable from its material cause. Similarly consciousness is separable from its known objects, but its known objects are not separate from consciousness. Therefore consciousness is the material cause of known objects and not vice versa. It follows that physical reality is a manifestation of consciousness, as a gold ring is a manifestation of gold.

It can be said that consciousness can be (can manifest itself as) several distinct things at the same time. When I look in front of me, several objects appear simultaneously. And after a moment still other objects manifest themselves to me. For example, if I look at the sky, the different clouds and also the blue background appear to me at the same time. After which, just a moment later, the clouds will have changed shape and position. Consciousness has the ability to manifest itself, that is, to be, as if it were all these objects, while remaining unchanged. No physical entity has the ability to manifest, that is, to be, multiple distinct things. Every physical entity is just that one thing that it is at any given moment.

Doesn't this ability of consciousness to be multiple things contradict the principle of noncontradiction? This principle says that an entity A cannot be A and non-A in the same sense as A and simultaneously. Consciousness cannot be (manifest as) cloud X and not be (not manifest as) cloud X (e.g. a different cloud Y) in the same sense as (i.e., without distinction/separation from) X and simultaneously. But consciousness perfectly respects this principle, and in fact the objects in consciousness are distinct/separate from each other: the clouds X and Y are perceived as two distinct clouds.

No entity (thing, process, phenomenon, concept, etc...) has the property of being able to manifest itself in more than one single way, only awareness-being has this ability. Therefore a computer, a brain, or any physical system not having this ability, can never be aware. Rather a computer, a brain, or any physical system are manifestations of, in and as awareness-being.

Once understood and thought through well, this argument becomes very intuitive. The reality we experience in the waking state is no different than that of the dream state: our avatar and physical surroundings exist within our consciousness. Everything we perceive has a profound unity.

9.

ARGUMENT FROM BECOMING

Being is and cannot not be, Non-Being is not and cannot be.
Parmenides of Elea

Physics is the study of the structure of consciousness.
The substance of the world is mental.
Sir Arthur Eddington

The ancient Greek and Roman philosophers, from Parmenides onwards, always found themselves in great difficulty in solving the problem of becoming, the problem of change.

The problem is that stating that an entity changes means that I first had a manifestation A, and then a different manifestation B. But this implies that A ceases to exist, while B begins to exist. But where does A go, and where does B come from? From nothing? But nothing, by definition, does not exist, and therefore nothing can come or go from nothing. Nothing Comes From Nothing: Ex Nihilo Nihil Fit.

Parmenides thought he could solve the problem by stating that: what is cannot not be, and what is not cannot be. How to disagree? On the other hand, it seems a formulation of the principle of noncontradiction, which is the basis of any discourse. Unfortunately this implies denying the existence of becoming. It is the solution of blindfolding oneself, or of saying: since I am unable to explain X (eg: becoming), then X does not exist.

In Western philosophy there have been several proposals to resolve this dilemma. The best known proposal is the Aristotelian theory of hylemorphism. This theory states that in every change there is an element that does not change, and an element that changes. The first is called (metaphysical) matter, not to be confused with physical matter. The second is called form. When I take a piece of clay in the shape of a horse and give it the shape of a turtle, I am taking the clay material and changing the horse shape to the turtle shape. The discourse can become increasingly sophisticated with the introduction of subtleties represented by the concepts of first matter and secondary matter, or substantial forms and accidental forms, synol and so on. The hylemorphism is just a conceptual chimera masquerading as a solution. For where does the shape of the horse disappear and where does the shape of the turtle come from? The problem is only moved, but not fixed. Certainly the scholastics would say that the form is not an entity, the entity is only the synol: that is, the union of matter and form; therefore it is not worth asking where the form comes or goes from. Clearly this is grappling with the appropriate language games.

Unfortunately, none of the proposals has proved to be fully satisfactory for solving the problem of becoming. For centuries philosophers have grappled with this dilemma, building huge conceptual castles in the air.

But in truth, as often happens, the solution has always been under our nose. In the end, what is the only "place" in which becoming manifests itself? What is the only "place" where any entity appears and disappears? Of course it is awareness. Whether or not there are entities out of awareness, it is a fact that it is in my awareness that the entities appear and disappear to me. So becoming is not about objects out there, out of consciousness coming into existence and ceasing to exist. Becoming is an experience inside my awareness. After that, only in a second moment, I can try to hypothesize that the experience of becoming can be explained by hypothesizing entities external to awareness. But this assumption is not only unnecessary but also contradictory.

9. Argument from Becoming

Let us recall the reason for the contradiction. Entities existing per se out of consciousness are contradictory with respect to becoming because they would imply passages from being to non-being and vice versa. First I do not have a certain substance A, then a certain substance A appears out of nothing, and finally a certain substance A disappears into nothingness. But since nothingness does not exist, then nothing can come or go from nothing.

The non-dual understanding of reality shows us that entities do not begin and cease to exist in themselves out there, otherwise there could be no becoming. Because the becoming of objects existing in themselves is contradictory. Entities begin and cease to exist in our awareness. Entities emerge and disappear in awareness. A bird begins to sing, and the sound begins to exist in my awareness. Then she stops singing and the sound disappears in my awareness. Entities exist in awareness, and awareness is the being in which entities exist.

In the non-dual perspective, becoming ceases to be a problem because when a manifestation begins or ends, from an absolute point of view, nothing changes, nothing is added or taken away. Before the manifestation there was only consciousness. During manifestation there is only consciousness, for manifestation is only consciousness. And finally after manifestation there continues to be only consciousness. Manifestations appear and disappear only from the relative point of view of some other manifestation. The only reality, the only being is awareness, which exists in itself, independently. Using the term of Vedantic philosophy: Consciousness is "satyam". On the other hand, the entities which have an existence dependent on awareness, which are neither real nor unreal, are called "mithya". The word mithya probably has the same origin as the English word myth, or the Italian "mito". This etymological aspect helps us a little to understand the meaning of the term mithya.

We can also coherently state that entities do not begin and cease to exist, but that they have always and forever existed. In fact, with respect to consciousness, one cannot say that an entity has begun to exist or that it has ceased to exist. Because consciousness has no reference system in which to locate the beginning and the end of the entity's existence. Any frame of reference is itself a manifestation in consciousness. An entity (a manifestation) begins and/or ceases to exist only in reference to another entity. Time is a classic entity that serves as a reference system for locating other manifestations. But other reference systems can also be imagined.

With respect to the various reference systems, entities can be manifested (therefore exist in a proper way) or be non-manifested. With respect to consciousness, entities simply are (what they are). Wanting to combine these two statements we could say that entities always exist:
1. Sometimes in a manifest way, that is, in an apparently dual way with respect to other entities. The entities are in fact separated from each other.
2. And "sometimes" unmanifest with respect to other entities. In this case we could say that they exist non-dually in consciousness: entities are in fact not separate from consciousness.

The term "sometimes" is not to be understood in a chronological sense, because time is also a manifestation. With the term "sometimes" I want here to convey the relative sense of dependent existence of each manifestation with respect to other manifestations.

10.

ARGUMENT FROM SUBSTANTIALITY

You will never change things by fighting against the existing reality. To change something, build a new model that will make the existing model obsolete.

Buckminster Fuller

There is no difference between our own being and the knowing of our own being.

Rupert Spira

This is the real secret of life: to be fully involved in what you are doing here and now. And instead of calling it work, realize it's a game.

Alan Watts

The Brihadaranyaka Upanishad (at passage 3.9.26) introduces the term neti neti, "not this and not that": "The Self is that which has been described as 'Not this, not that', it is imperceptible, because it is never perceived; incorruptible, because it never decays; detached, because it is never attacked; free, because it never suffers and never perishes". Neti neti is to be understood as the process of looking for oneself in oneself. Traditionally speaking, the first neti refers to denying that consciousness is something concrete, material, tangible: for example the body, the brain or a physical object in general. The second neti refers to the denial that consciousness is a mental phenomenon or something describable with a concept. This is because whatever we can experience, whether it is a concrete thing (the first neti) or conceptually describable (the second neti), cannot be who we we really are, the consciousness, since while we experience it there is always at the same time a subject who bears witness to these things. If we were one of those things, then there would be no testifying subject, because it would be on the other side as the witnessed object.

When meditating we try to bring attention to the thought or sensation of being I, we realize that all attributes, all objects, all mental contents disappear, and nothing remains. This direct experience shows us that there are no parts, no attributes, no contents in awareness.

If I affirm that "consciousness has no parts or attributes", one could object that this is just a sentence that appears to consciousness, but without real value, in short, that its negation could also have appeared, or nothing could have appeared; therefore it cannot be said with certainty whether consciousness has parts or not. It is like when one does a multiplication mentally (example: 42*42) and a correct result can appear mentally, as well as a false one. And feeling that the result is correct doesn't necessarily imply that it is.

This objection would be valid if we remain in the field of concepts: that is, of entities that refer to other entities. But consciousness is not known through a conceptual network, through inferences. On the contrary, it is known immediately, through direct experience: it is the knowledge of its own knowledge. Therefore the objection is not valid.

Awareness is therefore the ultimate subject: that which is devoid of objects. As the ultimate subject, if something appeared to awareness, then it would be an object in it. As the ultimate subject, consciousness cannot be explained by anything objective. Consciousness is therefore nothing deriving from physical reality.

Let's try to frame this argument a little more in depth. If a computer perform the neti neti procedure, there would be nothing left at the end. Neti neti is here to be understood as the process of looking for oneself in oneself.

10. Argument from Substantiality

If the computer perform neti neti, if it looked for itself, from an external objective point of view (as the electronic engineer would do) then it would eliminate all the hardware components one at a time until nothing was left. The computer might ask:
1. Am I the motherboard? No, the motherboard is just one of my components, but as a computer I am not limited to the motherboard alone. So let's exclude the motherboard.
2. Am I the external storage unit? No, then let's exclude the external storage unit.
3. Am I the monitor? No, then let's exclude the monitor.
4. Am I ...

At the end of the neti-neti of the computer nothing would be left.

If the computer perform the neti-neti, if it looked for itself, from a (pseudo-)subjective point of view, then a program inside the software would do it against the rest of the software programs, for example: antivirus, system update program, operating system, kernel or cpu resource management program, etc .... In that case it would exclude all the programs one by one until there is nothing left. The software would ask:
1. Am I the merge-sort function present in my program? No, the merge-sort function is just one of my components, but I as software don't limit myself to just the merge-sort function. So let's exclude the merge-sort.
2. Am I the hash table X present in my program? No, so let's exclude the hash table X.
3. Am I ...

In the end, again, there would be nothing left.

The neti-neti of a physical object is like Nagarjuna's chariot or Theseus' ship. Try removing one by one all components that are not the entity itself that you are looking for. Eventually, at the end of this process you will not find the entity you are looking for, you will not find anything at all, in fact there will be nothing left at all. Physical or abstract objects have no substance, they are not a substance, they do not exist in themselves, but only in relation to something other than themselves: their components, and who assembled them. Take away the forms and

81

eventually the names disappear as well. Things are just names and forms: nama-rupa.

A physical object is made up of parts and attributes because at the end of neti-neti there is nothing left that can be named as a such specific object. The parts and attributes of that object are in a sense that object.

But if consciousness does the neti-neti, if consciousness were looking for itself, the result would be totally different. At the end of the process, consciousness does not disappear: it would always remain present in all its entirety, without remaining deprived of anything and different from how it was before doing the neti neti.

Consciousness, therefore, has no names and forms; therefore it has neither parts nor attributes. While the computer objects (hardware components and programs) are external to each other and external to the computer (they are separated from each other); the objects appearing to consciousness are internal and not separate from consciousness.

But if consciousness has no parts, then how to understand things like the subconscious or the unconscious? Maybe they don't exist? The sub-conscious e the un-conscious are actually to be understood as sub-mind and un-mind respectively.

By doing the neti neti of consciousness one realizes that consciousness is not a nothing, it is not non-substantial like a physical object, it is not mithya, because at the end of the neti-neti remains identical to how it was at the beginning: the observer never disappears. Therefore, consciousness has no existence dependent on anything other than itself. Rather, it has a totally independent existence: it is satyam. Consciousness is not an object, consciousness is the ultimate subject, that is: is the ultimate substance.

Many of the logical and philosophical paradoxes derive from the assumption, often implicit, that things out there have a substantiality, that they are satyam. The famous Indian Buddhist philosopher Nagarjuna

10. Argument from Substantiality

said: All is clear to those for whom non-substantiality is clear. Nothing is clear to those for whom non-substantiality is not clear.

At this point we are able to explain the difference between "objective knowledge" and "subjective knowledge". Understanding this difference is essential to understand some passages of the next chapters.

An entity X "objectively knows" an entity Y, if Y can be separated from X. Conversely an entity X "subjectively knows" an entity Z, if Z cannot be separated from X: Z is X. This means that at the end of X's neti neti Z is encountered. But if X encounter its Z then X is a substance, and knows itself subjectively. But if, on the other hand, X at the end of the neti neti does not encounter its Z, then X is not a substance, and therefore cannot know itself subjectively.

The substantiality of consciousness implies that it can know itself subjectively, because at the end of the neti neti its substance remains. It is this substance that constitutes our subjectivity, our inner life, the awareness of being aware, being present to ourselves. Conversely, any physical object cannot know itself subjectively because it has no substance: it is like an empty box.

11.

ARGUMENT FROM ENTROPY

To know what you look like, you have to look in a mirror, but don't mistake that reflection for yourself. What is perceived by our senses and mind is never the truth. All visions are mental creations only; if you believe them, your progress stops. Ask for whom the visions occur, who is their witness. Free yourself from all thoughts, stay in pure awareness. Don't move from that.

Sri Ramana Maharshi

In the eyes of the imaginative man, nature is imagination itself.

William Blake

The second law of thermodynamics states that the entropy of a closed system always increases. Entropy can be seen as the degree of disorder of a system, i.e. the complexity of a system measured as the amount of information needed to describe it.

Suppose we have 100 coins placed on a table, all heads side up. To describe this configuration, a minimum amount of information is enough: it will suffice only to say that all the coins are all on one side. If instead we start flipping some coins then we are increasing the amount of information needed to describe the configuration. In fact we'll need to say something like: all coins are heads except the ones in the following list. The longer this list is, the more information we are providing. Suppose we consider as ordered a configuration with all the coins placed on the same side, while all the other possible configurations are

11. Argument from Entropy

considered as disordered. With 100 coins with 2 possible states (heads or tails) there are 2 to the power of 100 different possible configurations. So the ratio between ordered configuration and disordered configurations is 1 over 2 to the power of 100. The disordered configurations are extremely more numerous. Of course, we could assume, by convention, configurations with at least 90 coins resting in the same side as ordered. But the meaning of the discussion does not change: the number of ordered configurations is enormously lower than the number of disordered configurations.

Now suppose you start with an ordered setup, and then take all 100 coins and toss them in the air. When they land back on the table it is very likely that they will have a messy configuration, because there are so many more messy possible configurations than ordered ones. Coin tossing represents a generic transformation of a physical system from one configuration to another. The increase of entropy, of disorder, can therefore be reduced to a matter of probability.

If we wanted to decrease the entropy of the coins, we would need an external mechanism to turn them properly. It can be shown that by doing so we would be simultaneously increasing the entropy of the external mechanism.

The aspect that I am interested in underlining here is that: the increase in entropy is a matter of probability and it is a consequence of the fact that the entity under consideration is composed of more than one element. If you had only one coin, then entropy can't increase.

The phenomenon of entropy is one of the most fascinating and profound in nature and we could talk about it ad infinitum. In order to understand the importance and depth of the concept of entropy it is interesting to quote a famous observation by Sir Arthur Eddington: "The law that entropy always increases holds, I believe, the supreme position among the laws of nature. If someone points out to you that your favorite theory of the universe disagrees with Maxwell's equations, then so much the worse for Maxwell's equations. If it turns out to be contradicted by observation, well, these experimenters screw up

85

sometimes. But if your theory turns out to be against the second law of thermodynamics, I can't give you any hope; there is nothing to do but collapse in the deepest humiliation."

The second law of thermodynamics shows us how things never stay the same over time and that there is an inexorable degradation in things. This is how, for example, the aging of our body is explained. But the law of entropy is not limited to describing physical processes, but being in its essence a statistical law, it also describes the fate of organizations. If an organization is closed to outside influences, then over time its level of internal disorder increases. This translates into corruption, conflicts, etc … When a political class, a club, a group of friends, a family, become impermeable to external influences, they end up falling apart. When politicians feel they are no longer accountable to their electorate, either because they have established a dictatorship or because they think they are not controlled, then they end up corrupting themselves.

Even our personality when it closes like a hedgehog and establishes barriers to the outside in the form of prejudices, extreme shyness, lack of desire to go out and confront each other, etc... ends up impoverished more and more.

In nature there are no static things (understood as static entities), but there are only processes. A static thing is a process that has been photographed, ideally by our mind. Even processes are not things: in the sense that processes are also in continuous process, varying their parameters and their nature. A process that doesn't change like a static thing is just an abstract model in our minds. Any physical object is in constant flux, in constant transformation, panta rei. This flow is fractal: an organism is made up of continuously transforming organs; organs are made from ever-changing cells; cells are made up of molecules in continual transformation; molecules are made up of constantly changing atoms; and so on.

If we could film a mountain range for several million years and then watch the footage speed up a few orders of magnitude, we would

11. Argument from Entropy

see something like waves in a rough sea. By appropriately changing the time scale, all things are immediately revealed for what they are: processes subject to the law of entropy.

Every physical phenomenon is a type of process subject to increasing entropy.

Now let's look at any object that we have before our eyes, for example an apple. Let's start by pronouncing its name: "apple". After that we make sure that no more words, and therefore no concepts, appear. No concept of apples, fruits, greengrocer, etc... We look at the apple without conceptualizing anything. We then slowly shift our attention from the apple to the sense of being aware of the apple. Finally we shift attention from the sense of being aware of the apple to the sense of just being aware. To do this, we can try to arouse the sensation of wanting to look at the (imaginary) point from which we look at the apple. Obviously we won't find such a point, but we will get the effect of immersing ourselves more and more in the awareness of awareness. We stay there for at least a minute and then we resurface. As we resurface we notice that in this experience we have not perceived any change, no variation of anything, that time has seemed absent. Finally we note that this awareness, seen from the perspective of the mind, has been the same since we were born, that it has never changed, that I am always me, and I never stop being me.

If the reader has done this experiment correctly, it will immediately be clear that the law of entropy does not hold for consciousness. By repeating this experiment several times, you will notice that each time the experience remains the same. Awareness always remains identical to itself, our deepest sense of identity remains unchanged: there is no increase in entropy.

This implies that consciousness is made up of a single element (like a single coin): consciousness cannot be broken down into parts. Therefore consciousness transcends the realm of physical phenomena.

12.

ARGUMENT FROM PRESENCE

Things are not what they seem; nor are they otherwise.

Lankavatara Sutra

It is as if the Self were "created" in the process of self-reflection: I am because I know I exist, and I know I exist because I am (a slightly different version of Descartes' cogito ergo sum). I exist the instant I know I am, and I know I am because the "substance" I am made of is capable of self-reflection and in recognizing itself I become a Self. Self-knowledge is creative because it leads to the existence of the Self. Existence and knowledge are like two irreducible sides of the same coin.

Federico Faggin

You are not a drop in the ocean. You are the whole ocean in one drop.

Rumi

Once, when I was a child, my mother explained to me for the first time that my body had internal organs such as heart, brain, lungs. I remember perfectly how surprised I was, and asking myself: how come that I didn't already know? If I am my body then how come I didn't know how and what I am made of? Shouldn't I be "present" to myself?

These kind of questions have always kept coming up to me, and I've always found them very valid. The answer I have come up with is

12. Argument from Presence

that: I must certainly be present to myself, that is, I must be able to know subjectively: by intuitive, direct and immediate knowledge of what I am made of.

Consciousness is in fact self-reflective: it exists in the moment in which it knows (is aware of) itself in itself (it does not recognize itself in anything other than itself). This knowledge is different from all the others, because it is devoid of objective content and yet it is the most real and certain thing there is. Having no content then I cannot even put it into words. However, I can try to evoke it in the interlocutor by saying what it is not, with the process of progressive denial of neti neti. As explained in the argument of substantiality, at the end of the neti neti, what remains is consciousness. Once all the objective entities, all the contents (even that of the little inner voice) have been excluded, we realize that NOT nothing remains, but on the contrary something remains: the emptiness of the ultimate subject. Emptiness is not nothingness. Consciousness is this ultimate subject. Since this ultimate subject does not manifest any content, we still manage to "subjectively know" it as the constant "presence" in every experience, and this presence is our Self.

In this argument I want to first demonstrate that having a "subjective knowledge" implies knowing one's material cause. Next, I will point out that this material cause is pure consciousness, and from this it follows that consciousness is not explicable by anything physical, but is its own explanation. I recall that the distinction between "subjective knowledge" and "objective knowledge" was clarified at the end of the argument on substantiality.

If something, X, is not always present to myself then I am not X. Since my body is not present in the dream state, then I am not my body. Since I don't know exactly how my body is made up, then, again, I am not my body. Furthermore, since the mind is the sum of my thoughts, emotions, sensations, and that all of them are not always present, then I am not my mind either.

However, two objections arise to this argument. The first is that there are mental states in which I don't seem to be present to myself. For example in the dreamless sleep state it would appear that I am not conscious, and therefore not present to myself. The second objection is that it is necessary to clearly define what is meant by being present to oneself. There is probably no reason to think that we need to keep in mind what we are all made of. For example, if I am a brain process, then only the process and not the underlying brain must be present to me. As if to say that: a software knows the data it processes, but not the hardware in which it processes them. Let us analyze these two objections carefully.

Dreamless sleep is not absence of consciousness, but consciousness of absence. Dreamless sleep is simply characterized by the fact that there is no content present in consciousness, including time and space. In an experience without content there is nothing to remember, because remembering is always about something. Therefore we have no memories of dreamless sleep even though consciousness is present.

Either we say we experience dreamless sleep, even though we have no memories of it, or we don't experience it. If we experience it then consciousness is present (since consciousness is what it knows in every experience). If dreamless sleep were not our experience, then we would have to question the presence or reality of dreamless sleep itself: precisely because we have no experience of it. How else, other than experiencing it, could we notice that in dreamless sleep there was no experience? To notice something, that something must first be in our consciousness.

The experience of absence means that there are no objects present in consciousness. Time and space are also objects in consciousness. Therefore the dreamless sleep state is a "state" only when viewed from the point of view of the waking state. In fact, a state is that which has space-time references. But if in the dreamless sleep there is no time and no space, then it is not a state. I can say that I was in the "state" of dreamless sleep in my room between 3 and 4 am, only in the spatio-

12. Argument from Presence

temporal reference system of the waking state. But in the reality of dreamless sleep, the experience has no time duration or location.

In the state of dream, on the other hand, space-time references appear, which are quite different from those of the waking state. In particular, the space of the dream state is completely uncorrelated with that of the waking state. I can dream of an immense space, for example the panorama from the top of the Everest, yet my waking state body is locked in my room. On the other hand, the time of the dream state could be partially correlated with that of the waking state, although the durations may be completely different. A dream that lasted a few minutes in the waking state may have lasted a lifetime in the time frame of the dream.

We now come to the second objection. If by some chance we were to totally lose the use of the senses (sight, hearing, touch, taste, smell), and thus be unable to get any sensory input from the physical world, then the only things we would be aware of would be our mental images, thoughts and emotions. In reality this is what happens if you are inside a sensory deprivation chamber.

If we suddenly found ourselves in such a situation we could never say with complete certainty whether we were dreaming or awake: the two states would be indistinguishable. In fact, suppose we enter that state while awake and then fall asleep and wake up. Would anything change when you wake up? Nothing would change: because the contents of our consciousness would always be an internal product of our mind. Suppose that as soon as we were born, we entered this state, then, theoretically, we would never know that we had a body and a brain: these would not be present to us. So what will we think we are made of?

Based on what has been established in the previous chapters, we know how to answer this question by saying that what we are made of is consciousness. But let us try to reach this conclusion by another route.

We say that a physical system A (e.g. a thermometer) knows (at least from an engineering point of view) a physical system B (e.g.: the temperature of a room), when the state of A is influenced by the state of

B. A thermometer knows the temperature of the room because the latter influences the thermometer through, for example, the expansion of a fluid contained in it. If a fire breaks out in the room then the thermometer would register a very high temperature. The thermometer "knows" the existence of the fire thanks to the fact that it influences the state of the thermometer. If the thermometer were broken, then it wouldn't detect the fire. For the thermometer, fire exists if it knows it, in physical terms: if it can be influenced by it. The fire is the explanation, or cause, for the thermometer registering a high temperature.

We say that an entity Y knows an entity X when the state of Y is a function of the state of X. The state of X explains and/or causes the state of Y. In this context we define the "state" of Y, or X, in the most general way possible, that is, like any positive statement that can be made about Y, or X. Equivalently stated: any non-negative description of an entity is its state. If, by removing isolation of Y from X, this entails having to change the description of Y, then we say equivalently that:
- X is an explanation (or cause) of Y;
- X exists according to Y;
- Y knows X;
- X is present at Y;
- X and Y are not isolated.

Causation, explanation, knowledge, presence and isolation are all aspects (or names) of a single phenomenon. An appropriate variant of the concept of information could also be introduced, but I believe that the concepts of "isolation" and "presence" make the continuation of the argument more intuitive. In this approach (of a very engineering mentality) every knowledge implies a form of causation and vice versa.

Moreover, an immediate corollary of this approach is that an entity cannot remain isolated from the set of all its possible explanations/causes. In fact, remaining isolated would mean having no explanation, therefore non existence.

If I describe a room, X, as cold, then I expect that the description of the thermometer, Y, therein, will include that it marks a low temperature. The description of Y depends on the description of X. In

12. Argument from Presence

physics a description is generally represented by the value of a suitable physical dimension.

If I move the thermometer out of the room, the description of the room will not change, but that of the thermometer will change. So the description of the thermometer depends on that of the room but not vice versa. When I isolate X and Y from each other, and note that only Y changes description, then it means that it is X that explains Y, and Y depends on X. In particular, I define X as an efficient cause of Y.

If I have two thermometers, Y1 and Y2, in the same room, I expect them to read the same temperature (with some margin of variation). But if I put Y1 and Y2 in two different rooms, I will find that both change the temperature. Therefore if by isolating Y1 and Y2 both change description, then I deduce that there was a correlation but not a causal relationship between them.

If from the thermometer, Y, I disassemble the sensor, Z, I expect the description of Y to change dramatically. The thermometer will no longer read any temperature, and must be described as not working (or as the case may be as broken or no longer existing as thermometer). In this case not only one aspect of the description is changing, but the "type of description" is changing, so that Y can no longer be called Y. Y ceases to exist as Y. A thermometer without a sensor cannot be called a thermometer. We will be able to say that Y, as such, cannot be isolated from Z. If Y were isolated from Z it would no longer be Y, it would be a W. I define Z as a material cause of Y.

It then turns out that the following three statements are equivalent:
1. the material cause cannot be isolated from its effect;
2. the material cause is always present in its effect;
3. the effect always knows its material cause.

We have thus succeeded in giving a precise meaning to the concepts of material and efficient cause. Let's try to summarize. If isolating X from Y changes the description of Y and that of X does not change, then X is an efficient cause of Y. If I cannot isolate Y from Z, then Z is a material cause of Y. We could say that the efficient cause is a weak version of the material cause.

Let's go one step further and consider an entity O that knows Y. The state of O changes as a function of the state of Y. If Y is not isolated from Z then O will observe a normal Y. If instead Y is isolated from Z, then O will observe something different: a W. If I (understood as O) observe the thermometer (understood as Y) equipped with a sensor (understood as Z), I will acquire a different knowledge from the case in which the sensor (Z) is isolated (disassembled) from the thermometer (Y). Thus O's knowledge of Y cannot be isolated from Z. Since O changes state based on what it observes then it is changing its state based on whether or not Z is present in Y. Thus an observer O when knows an object Y, it is also knowing the material causes of Y, i.e. what cannot be isolated from Y. Therefore, knowing an entity means knowing what that entity is made of. Knowing has here to be intended in a broad sense, not only just conceptually, but as synonymous of being influenced.

Let's think about when we receive a well-wrapped gift in a box. Let's suppose that whoever gave us the gift made enough effort to choose the gift that we don't have the slightest idea of what's inside the box. Before opening the box, the only thing we can say is that there is a "gift" inside. However, the word "gift" is only a placeholder of the actual content, in itself it is not a real knowledge. I will gain a real knowledge of the contents only when I open the box and see what it is made of. Effective knowledge of the gift cannot be isolated from knowledge of what the gift is made of, from what the gift cannot be isolated from.

But could it be that consciousness's knowledge of itself is just the box, and not its contents? It can't be because the box is just the name that acts as a placeholder for the content. As a placeholder it is virtually isolated from any possible content: anything can be inside the box. Instead, as we well know by now, with consciousness we find ourselves in the opposite case: consciousness is isolated from everything; we can say what consciousness is not, but not what it is. The box of consciousness is open … and it is empty …. Consciousness is emptiness because it is the unmanifest.

However, O's knowledge of Z is not complete, only Y knows Z completely. If O fully knew the Zs of Y then it would be indistinguishable from Y: O would be Y. By incomplete knowledge I

12. Argument from Presence

mean that the state of O is not sensitive to any influence from the state of Y and/or Z. A camera is sensitive up to a certain number of pixels: I can't "see" the atoms with a reflex camera. A camera is only sensitive up to a certain number of frames per second: I can't track a bullet with a normal camera. If I look at a physical object with my naked eyes I cannot see that it is made up of atoms. The description of O does not change based on any change in the description of Z. Looking at a steel object with the naked eye, I cannot describe what its percentage of carbon is. On the contrary the description of Y varies according to the change of any description of Z. If the steel object changed the percentage of carbon then the description of its mechanical characteristics would change.

If I observe a book I am automatically observing, at least in part, what the book is made of: the cover, the paper, the ink. If I isolate these three components then I have nothing left of the book to observe. What cannot be isolated from an entity also cannot be totally isolated from the knowledge one can have of that entity.

Since consciousness (understood as O) knows itself (understood as Y) then it must know what it is made of, its material cause (understood as Z). But the only Z constantly known to consciousness is exclusively consciousness itself: its own presence. Hence consciousness is its own and only material cause. Among other things, given that in this case O coincides with Y, then consciousness must totally know itself.

Let's make this discussion a little more explicit. The only positive statements we can make about consciousness concern exclusively its objective content, and not consciousness itself. In fact we have seen, and we will go deeper into it later, that consciousness is in itself ineffable: we cannot say anything positive about it: we cannot say what it is. We can only go the negative way of neti-neti: saying what it isn't. We can only describe the content of consciousness but not consciousness itself. However we have established that any material cause of consciousness should always be known, should always be present to consciousness. However, there is no objective content that is always present in consciousness. This implies that consciousness can be isolated from its

95

objective contents, therefore consciousness has as its material cause nothing that appears to it as content, nothing that is objective. The only entity ever present to consciousness is consciousness itself. Therefore consciousness is its own explanation.

In addition to this we also note that the different states of consciousness (wake, dream, sleep, trance, deep meditation, coma, etc...) can in some cases be completely isolated from each other. For example, when we are dreaming there are no physical entities: dreamed lights, sounds, smells and tastes are not representations of anything physical.

It may happen that while we sleep our physical body is thirsty, and then we dream that we are thirsty and maybe we wake up and drink: the two sensations of thirst (wake and dream) are correlated and are of the same type but they are not the same instance.

The thirsty feeling of a dream cannot be quenched by physical water, by having drunk just before going to sleep. And waking thirst cannot be quenched by dreaming of drinking.

Also the dream state is not like a movie or a video game. If my avatar in a video game is thirsty, his thirst will be quenched by virtual water from the video game, and my avatar will stop needing water. But I, the player, don't feel thirsty together with my avatar.

Dreaming and waking can therefore be isolated from each other. Therefore some contents of consciousness can be isolated from the entities present in the waking state: the latter are not the material causes of all the contents of consciousness. A fortiori, consciousness cannot have physical entities for material causes, as these are present only in the waking state. In general, consciousness can be isolated from the entities present in all its various states (wakefulness, sleep, hallucination, trance, etc...): consciousness can be isolated from its contents. Therefore the causes or explanations of consciousness are not to be sought in its contents.

A thought does not think, the thinker is consciousness: the contents of consciousness are not present to themselves, but are present to consciousness. Therefore we observe that the contents of

12. Argument from Presence

consciousness cannot be isolated from consciousness. Ergo, consciousness is the material cause of its contents: it is what the contents are made of.

Once we understand that stating that: "consciousness is conscious of itself" implies stating that: "consciousness is conscious of its material cause", we understand that we can do a simple and rapid experiential search for the material cause of consciousness. What is that unique entity that is always present in any experience, whether in waking or sleeping, in meditation, etc…? What is that unique entity which cannot be isolated from any experience and which is therefore its material cause? It is consciousness itself.

Let's make one further consideration. How do I know that last night's dream wasn't real? If we think about it carefully, the only reason we can propose is that the dream appears unreal once we wake up. From within the perspective of the dream everything appears perfectly real and coherent. Only in the waking perspective does the dream appear unreal to us and often even absurd. Generalizing the discussion, nothing prevents waking from being like a dream, which appears real as long as we are in it, while it appears unreal when we wake up in another state of consciousness. Well, this is exactly what many of the mystics of all ages and cultures have testified to. Then, we realize that, in general: a reality (dream, wakefulness, etc...) appears real to us from within, and appears unreal to us from the outside.

But is there anything that is always truly real, regardless of perspective? Yes, it is consciousness. Consciousness is the only entity that is always "present" and immutable in any perspective, in any reality. Consciousness is real in all realities. Consciousness is called in Sanskrit satyam: always real. While the other realities, being real only from within, are called mithya in Sanskrit.

Let us consider a possible objection. In substrate independent processes we can describe the process without mentioning the substrate. Let's take the emblematic example of software: this is independent of the hardware it runs on; the software can run on different hardware and be described without reference to the hardware architecture. Intuitively it would seem that the hardware is the material cause of software, but

97

the former is not included in the description of the latter. The individual functions of the software can be fully described without reference to the hardware. Similarly, consciousness may be a substrate-independent process and therefore its contents may not be described using waking physical entities. And yet as beneath software there is an hardware, so also beneath consciousness there may be a material physical cause.

This objection loses its meaning as soon as we realize that the description of a piece of software deliberately employs only efficient causes. Because in designing software we are not interested in material causes. Substrate independence is only a conceptual isolation, and is functional to compensate for the human incapacity to program in machine language. In object-oriented programming terms, software is a set of objects that change the state of other objects. Software objects are related to each other only by efficient causality. Each object can be isolated from the others, and therefore does not constitute the material cause of other objects.

To trace the material cause of a software object, a function, or any software entity we must ask what is describing what, and iterate this question up to a level where conceptual isolation can no longer be applied. A software is a convenient mental description of what the hardware does. In the same way that a mathematical model is a convenient mental description of what a physical system does. But neither the software nor the mathematical model exist as physical entities: they are only mental entities. Since it is very difficult for the human mind to program in machine language, then let's create appropriate higher-level descriptions in the form of programming languages easily managed by our brain. If I find a function in software that, for example, performs a summation, I have to ask myself: what does this function describe? Eventually it will describe a set of simpler functions in a lower level language. Then I ask myself: what do these new single functions describe? After a couple of iterations I get to describe the behavior of a group of hardware transistors. Transistors are the material cause of software.

12. Argument from Presence

It follows that independence from the substrate is only conceptual: it is a mental abstraction. Given a description in terms of efficient causes, I can always go back to material causes by asking: what is being described. Therefore the objection is not valid.

13.

ARGUMENT FROM INEFFABILITY

Your true nature is indescribable. It cannot be known with the mind, yet exists. It is the source of everything.

Sri Nisargadatta Maharaj

Our normal waking consciousness... is but a special kind of consciousness, while all around it, separated from it by finer screens, lie entirely different potential forms of consciousness... No account of the universe in its totality can be definitive if these other forms of consciousness are not taken into account.

William James

The perceived cannot perceive.

Huang Po (Chan/Zen Buddhism Master)

This argument is based on the ineffability of awareness, that is, on the fact that it is not describable and not explainable objectively. Awareness is not describable and explainable because whatever image or concept one uses to describe and explain it is itself an object in awareness, and therefore the latter is distinct from the former.

Why would it be distinct? In Advaita Vedanta there is an investigative principle called DṛgDṛśya-Viveka, which means: that which sees is distinct from that which is seen. DṛgDṛśya-Viveka is an expression of the more general negative path expressed as: the knowing subject

13. Argument from Ineffability

cannot be the known object. I have elaborated three different lines of argument to demonstrate that the perceived cannot perceive. Let's examine them one by one.

The first line of argument is the following. If the subject could be an object of knowledge, who would know it as a subject? If I objectively knew the consciousness, the ego, the subject, then automatically there would be a second subject who is knowing the first. In turn, this second subject is known by a third. And so on in an infinite regress. But our direct experience shows us that the subject is not known by a second subject, but on the contrary, it knows itself: it knows itself subjectively and not objectively. This means that consciousness is not like a physical system in which one part knows another. Consciousness in its entirety knows itself entirely. The conscious subject is self-luminous: he knows itself by itself and in itself. Consciousness is the ultimate self-knowing subject.

Pay attention, consciousness does not know itself objectively, on the contrary consciousness knows itself subjectively: it knows itself by being itself. This point was made clear in the argument from presence and at the end of the argument from substantiality. In this argument I want to demonstrate that anything objectively known cannot be consciousness.

Suppose I'm in the theater, and I'm the only spectator who watches the actors perform on stage: if at a certain point I go on stage to act, who's left to watch the theater work? Nobody. Now since awareness is never absent in any experience, then, metaphorically speaking, awareness never stops being a spectator and never goes on stage to become an actor. Awareness is not anything known.

In the argument from asymmetry we have clarified the difference between being present "on" (or "to") and being present "in" consciousness or another entity. When an image X appears on a computer screen, we say that X is present "on" the screen but not present intensively "in" the screen. That is, the image is not inside the screen: the screen is not made of the image and/or the image is not made by the screen. The content of the screen image tells us nothing about the hardware of the screen. On the screen, as in all physical

101

systems, there is a difference between appearing in front of (on-to) and appearing inside (in). Conversely, what appears to consciousness also appears in consciousness, and vice versa. In consciousness there is no difference between a front and an inside. Everything inside is also in front and vice versa. This implies that if consciousness has a property X (X is inside consciousness), that property must also be known by consciousness (X is in front of consciousness). Consciousness is self transparent. The being of consciousness is its own knowing itself. If I want to know what the screen consists of, how it is made, I have to disassemble it and look inside, and discover the electronic circuits, the liquid crystals, etc...: I can't limit myself to looking in front of it, at its surface. Instead, if I want to look inside consciousness, to find out what it consists of, I don't have to disassemble it, because what is inside it is what is in front of it. So let's go and see what consciousness consists of, checking what we find inside it. Inside, we will find the known objects, but not only. As seen in the argument from substantiality, in the presence of any object (concrete or abstract) we can identify the conscious subject as the conscious space in which the known object is immersed (present): like the moon is immersed in physical space. Together with each physical object, the physical space in which it is immersed is always present, and space and object are distinct and distinguishable. Similarly, together with every known object, the conscious space in which it is immersed (in which it presents itself) is always present. The ultimate subject is precisely this conscious space, which is therefore distinct and distinguishable with respect to the known objects, and being also transparent, I can then look inside. Looking into it, I don't see anything objective inside it, therefore it turns out that this conscious space is also separate, further, with respect to known objects. The transparency of consciousness implies that in it: being distinct (gnoseological relationship) and being separate (ontological relationship) with the known objects coincide.

This means that consciousness, the ultimate subject, is distinct. -separate from whatever is present in front of it. While the latter is not distinct from consciousness.

So, every time we believe we have succeeded in objectifying consciousness as a certain object X (a process, a physical body, a set of attributes, an idea, a mathematical model, etc...) we must ask ourselves:

13. Argument from Ineffability

in which space is X known? In which space is X present? Then we understand that this conscious space always remains distinct and beyond X, unattainable, indescribable, indefinable: therefore the conscious subject is always distinct-separate from the known object, and therefore cannot be described in terms of the latter. What perceives is not perceptible, the perceptible cannot perceive.

At this point an objection could arise: it is understandable that consciousness is not the object, X, that is knowing, in fact objective knowledge is such precisely because the known object is somehow distinguishable or separable from the subject. However, X could have the same structure (form, image, model or functionality) of a component, partial or total, of consciousness. A bit like the image of our face in the mirror. The image is not our face, yet it has the same structure of the face, so we can say that by knowing the image in the mirror we indirectly know our face.

However, the image of the face is not known by the face, but by the mind. It is our mind that, through the image on the mirror, knows the structure of our face. In turn the mind is not known by the mind, but by consciousness. Nothing can know itself objectively. Only an entity, Y, other than X, could know X objectively.

Now suppose we know an object, X, which has the same structure, S, of a component, C, of consciousness. X could be a conceptual description (in physical or non-physical terms) of consciousness. By definition, S is what X and C have in common, in the sense that there is nothing that can distinguish S in X and C. S is a single, unique and indivisible conceptual entity. Now, objectively knowing X does not imply objectively knowing S as well, and through this objectively knowing consciousness. In fact, if I knew S objectively then I would be splitting it into two distinguishable entities: an S object and an S component; but this contradicts the definition of S. Therefore S cannot be known objectively. Therefore consciousness cannot be known objectively: consciousness is ineffable, it cannot be describable conceptually.

The second line of argument is as follows. The object causes an effect in the subject, and this effect is contained in the subject and it is the knowledge that the subject has of the object. The subject cannot cause an effect on itself, and it cannot contain itself, unless it has to separate itself from itself by subdividing itself into parts (concrete or abstract); such as one part containing and one part contained, or one part cause and the other effect.

When we say that a physical system knows itself, we are speaking improperly. What happens is that one part knows (=physically interacts) with another part. In one way or another the subject always remains separate from the object. In fact, in a cognitive process what happens is that the object causes, directly or indirectly, an effect on the subject. And this effect is the subject's knowledge of the object. Now an entity can cause an effect in itself only in the sense that one part acts on another part. Scientific reductionism works thanks to this principle: to explain an entity (an organism, a society, an atom, a machine, a molecule, a cell, a planet, etc...) in terms of the interactions of its parts. If subject and object coincided, there could be no effect and therefore no objective knowledge.

Without going into detail about the different ways in which a cognitive process can take place, we can in general affirm that: as long as there is objective knowledge, the subject is distinct from the object: the knower is distinct from the known, the this of the subject is distinct from the that of the object. If the distinction fails then the cognitive process fails. But where we can speak of objective knowing, then the knower remains separate from the known.

If I place a video camera in front of a mirror, it will have a sort of knowledge of a part of itself: its lens knows the outer envelope. If, on the other hand, I want the camera to specifically know its lens, then I disassemble it and place it in front of the camera. Unfortunately what will happen will be that the video camera will stop working and there will no longer be a cognitive process. The lens cannot be objectified by the camera: the camera is unable to maintain the subject-object distinction between itself and its lens. The camera, as a camera (therefore working), cannot know its lens. The camera cannot objectively know what cannot be separated from itself.

13. Argument from Ineffability

Any physical entity, X, cannot objectively know what cannot be separated from itself, because it is unable to cause an effect on itself. X cannot separate itself from itself, remaining itself, and therefore cannot objectively know itself. If, on the other hand, X were properly subdivided into distinct parts, then one part could know another. But in that case it is no longer X in its entirety that knows X in its entirety. The this of the subject cannot become the that of the object, remaining subject to the effect (=knowledge) of the object.

Now we know for sure that where there is consciousness there is self knowledge: consciousness exists as long as it knows it exists; consciousness and awareness are synonymous. Consciousness is subsisting knowledge. Consciousness is knowledge that exists in and for itself. Therefore where there is consciousness there is also always the distinction between subject and object.

Everything can be the object of knowledge by the consciousness, except the ultimate subject: itself.
Therefore consciousness, as the ultimate subject, is nothing that can be known objectively. Consciousness is separate from all that is not itself, the ultimate subject.

It follows that any materialistic theory, or otherwise, aimed at objectively explaining awareness, that is: based on anything other than awareness itself, loses value. Because the concepts he would use, as known objects, would refer to entities with respect to which awareness is distinct. So instead of being a theory about awareness, it would be a theory about those objects mistaken for awareness.
If I can know something then I cannot be that something.

With appropriate medical studies I get to know the physical and functional characteristics of my brain. These characteristics become, thus, object of knowledge by my consciousness. As objects of knowledge they are separate from my subject-consciousness, and therefore my consciousness is not a function of my brain. Let's analyze the argument in more detail. We are saying that if consciousness were a specific function, X, of the brain then it could not know it. In fact, for consciousness to know X, it is necessary that X causes an effect (knowledge) on itself (consciousness). But we have established that this is not possible. Therefore any function of the brain knowable by consciousness is not consciousness. Consciousness is potentially able to know all the functions of the brain other than the specific one in which it eventually consists. However, considering that there are no theoretical reasons which prevent us (one day) from knowing all the functions of the brain, it follows that consciousness is not any function of the brain.

If a scanner scanned the drawing of its operating mechanism, in which all its components are represented, would it be knowing itself? And thus violating the separation of subject and object? No, because what it would know is just the shades of gray of the ink on the paper, and the scanner is not a set of shades of gray.

And if instead an artificial intelligence, AI, were shown its algorithm, would it be knowing itself? And thus violating the separation of subject and object? No, because if one analyze this case carefully, one would notice that what happens is that one part of the AI knows another part of the AI. The parts of the AI are the functions and data that compose it. It is never the case that the entire AI knows itself. As explained in the argument from Asymmetry, each physical entity is separate in itself, it is granular. So every physical entity can know itself only in the sense that one part knows (= interacts) with another part; therefore the separation of subject and object is respected.

I am not what I can objectively know. I may have sadness, but I am not sadness. But how to realize it also experientially? A transparent glass placed in front of a red object apparently takes on the red colour. To be absolutely sure that the glass is not red, it is enough for me to place it in front of a non-red object, and to see that it will apparently take on a different colour. Similarly, if I succeeded with a particular technique in emptying consciousness of its contents for a brief moment, I would realize from direct experience that it is nothing that can be objectively known. This technique is meditation. For example, I can meditate by repeating a mantra over and over and focus on the silence between one repetition and another. Slowly I increase the duration of this interval of silence. For many people (but not for all) during that silent

13. Argument from Ineffability

interval it happens that the consciousness is sufficiently emptied of mental contents. At that point one became aware that one cannot be any objective content.

This very simple argument on ineffability has hitherto exploited the negative path of the distinction (or separation) of consciousness with respect to its contents. However, ineffability can also be argued in the opposite way: by exploiting the positive path of non-separation of contents with respect to consciousness. Let us now try to develop this second modality, in the following third line of argument.

If concepts appear in consciousness, then how do they speak to and refer to consciousness? The concept of consciousness is an appearance in consciousness. It's not consciousness. But then what does the concept of consciousness refer to? Isn't it that we're talking in vain?

Well, consciousness is separate from the concept of consciousness. Consciousness itself is ineffable. However, the concept of consciousness is not separate from consciousness, it is made of consciousness, and refers to consciousness because it is made of it and appears and disappears in it. The case of the concept of, for example, a tree is different. The concept of trees does not appear in trees and is not made of tree. Consciousness is self-luminous: in every experience it reflects itself, it is aware of itself. For since everything is made of consciousness, it follows that consciousness sees itself in everything. Knowledge of everything is always accompanied by an awareness of the Self, because everything is made of consciousness, of the Self.

Someone could object that all this discourse is tautological: clearly everything is apparently inside the consciousness because, by definition, consciousness only knows what it knows; what he does not know is out of consciousness and nevertheless exists.

Let us elaborate a little on this objection. Let's imagine that the brain has a cognitive faculty called the mind. This mind registers signals from the outside world and processes them. This processing is characterized by a particular quantitative property X. X could, for

107

example, be all the higher the more the information it produces in output is integrated with each other. The X is not uniform throughout the mind, there are parts with a lower X, which we will call unconscious parts, U, and then there are parts with a higher X, which we will call conscious, C. C and U are not to be understood as much as static things, rather as information processes. The different parts of this mind are more or less integrated with each other. In fact, the different parts, whether conscious or unconscious, exchange information with each other. Of course, only the information found in the conscious parts is experienced consciously. By definition, I find no information in the conscious part that is not consciously experienced, and when new information N enters from an unconscious part, N is immediately experienced consciously. Therefore everything is apparently inside consciousness, C, because C does not access U, but U exists and exchanges information X with C, and C is not aware of this exchange: C is not aware of the meta information, MX, on the information exchange of X.

Visually we could imagine this mind as a sea. This sea is conscious of the fish floating on its surface, while it is unconscious of the fish remaining submerged. When a fish moves to the surface the sea becomes aware of it. After which, when the fish plunges again, the sea ceases to be aware of it.

Another physical analogy can be that of the change of state between fluid and solid. The mind is like the arctic sea where the surface is solid and underneath it is liquid. Following the analogy, the Arctic Sea is aware of the ice on the surface, and unconscious of the liquid below the surface. An underwater current that emerges on the surface became solid and the sea becomes aware of it. Then, during the summer, part of the ice on the surface melts and the sea becomes unconscious.

This objection immediately falls on the basis of everything we have already established in the previous arguments. The explanation and analogies might be valid if we are talking about the mind, but not about consciousness. Indeed, we have established that when we are aware of an object, that object is not separate from consciousness. This implies that

13. Argument from Ineffability

the substance of the object in consciousness is only consciousness itself, and cannot be decomposed into a more fundamental substance. The existence of the object in consciousness is entirely dependent on consciousness. There are no unperceived perceptions. A feeling exists only when it is perceived, if I stop perceiving it then it ceases to exist. Everything that makes up the object in consciousness (a feeling, an image, a sound, etc ...) is made up of consciousness. The substance of the object in consciousness is wholly constituted by consciousness, and consciousness in turn is not constituted by a more fundamental component (perseity). Therefore this substance cannot change state: transform itself from unconscious to conscious and vice versa.

The ocean current analogy is invalid because ice (which according to the analogy represents a conscious experience) is reducible to a more fundamental component: water. Ice is not defined per se: it has no perseity. Ice is defined by water: it is in fact defined as a phase of water. Conversely, consciousness is not composed of a more fundamental component. Therefore the analogy is not valid.

The fish analogy is invalid because fish (which according to the analogy represents conscious experience) are not made of surface seawater (which according to the analogy represents awareness), but of organic cells. Conversely, an object in consciousness is made of consciousness. Therefore this analogy is also invalid.

In general: an entity A can be transformed into an entity B, only if the components that make up B are also found in A. If B has a component that is not found in A, then it cannot be produced starting from A. If, a fortiori, B consisted of a single component X, and the latter is not found in A, then one cannot transform A into B.

In consciousness there is one and only one substance, consciousness itself, which manifests itself by appearing as if it were different objects. Therefore the correct analogy, but up to a certain point, to visualize the relationship between consciousness and the perceived objects, is that of an object of pure gold. A bracelet made entirely of gold can be transformed into a gold ring, but it cannot be transformed

109

into a silver ring. Because the substance of which the bracelet is made is exclusively gold, and there is not a single gram of silver. An object in consciousness is made entirely of consciousness/gold, and therefore cannot be transformed into unconscious/silver. Same the other way around.

Therefore, if one has a flow (of information or other) inward and/or outward of consciousness, then the substance of that flow is not consciousness. It follows that there cannot exist a flow of objects (of information) between consciousness and an entity external to consciousness. The objection is therefore invalid.

Going round all the phenomena, from the most concrete to the most abstract, one discovers that they are all in consciousness and made of consciousness. This is an empirical finding. Any conceptual discussion that tries to demonstrate a world external to consciousness is also globally, as an explanation, and locally, with the individual concepts it employs, entirely within consciousness. It is empirically ascertained that everything is within the consciousness and there is no world out there.

Whatever definition or explanation one tries to give of consciousness is incorrect: because it is an apparition in consciousness and therefore it is not consciousness itself. Anything perceived cannot perceive, anything known cannot know. I can think of a whole range of hypotheses, for example that consciousness is …

1. … integrated information;
2. … an emerging phenomenon;
3. … a computation;
4. … a narrative center of gravity;
5. … a phenomenon related to quantum mechanics;
6. … a "strange loop";
7. … an illusion, in reality it doesn't exist;
8. … the common working space of the cerebral faculties;
9. … a self-organizing phenomenon, such as swarm intelligence;
10. … the mind;
11. … a fiction, useful for the purposes of natural or artificial selection;
12. … a particular neural loop;

13. Argument from Ineffability

13. ...

All these definitions and explanations are incorrect due to the simple fact that all they are, use and refer to are concepts, ideas, images, patterns, structures, models, abstractions, modalities, etc ... which exist only as objects in consciousness, of consciousness and are made of consciousness. Consciousness is always beyond any manifestation. Consciousness itself is ineffable.

Awareness is ineffable: all physical and mental objects appear in consciousness. Any description of consciousness (including the hypothesis that it is something physical) is an appearance in consciousness and does not exist out there per se. Therefore the only possible explanation for consciousness lies in consciousness itself, consciousness is not mithya, consciousness is satyam.

14.

ARGUMENT FROM CONCEPTS

Love says: "I am everything". Wisdom says: "I am nothing". Between the two, my life flows. Since at any point in time and space I can be both the subject and the object of experience, I express this by saying that I am both and neither, and beyond both.

Sri Nisargadatta Maharaj

Confusing names with nature, one ends up believing that having a separate name means being separate. And this is the spell.

Alan Watts

We define as concepts those objects known to consciousness that are "conceived" within the mind and are experienced with the use of words. Words can present themselves as a visual image or as a sound in our mind. The intentionality or meaning of a word is explained by the fact that by pronouncing it or seeing it written, a whole series of other mental images are perceived in a more or less mixed way: visual, sound, tactile, etc…

We therefore have a word P perceived in the foreground plus a set of other images X (of different types: visual, sound, etc…) perceived in the background. Given a concept C, we have the feeling that its P refers to something of which the concept is a concept, but this something is nothing other than X.

14. Argument from Concepts

Let's take the word/concept of "airplane". When we mentally pronounce the word "airplane" we will have:
- P: the "airplane" mental sound, well in the foreground, perfectly distinguishable.
- X: various visual images of airplanes; a sample of our tactile sensations when we are sitting on an airplane, a sample of airplane noises, etc... then maybe also sensations related to that of an airplane such as: airport, suitcases, travel, etc... All these Xs are each placed at different degrees in the background of our attention. X also encompasses the various connotations of the concept.

The moment an X moves from the second plane of attention to the foreground, then it becomes the next thought. If I read the word "red" (thought 1) it is possible that immediately afterwards I have a concrete mental image of red (thought 2). Thought 1 was a P, which had many examples of red indistinctly in the background, Xs. After which one of these examples moved into the foreground becoming the second thought.

Every concept has this double aspect: primary and secondary. Even the concept of "atom" has this double aspect:
- P: the sound (or written) word "atom".
- X: various graphic images of the atomic model, the periodic table of elements, related concepts such as electron and proton, the image of the book or of the classroom where we learned this concept; the name Rutherford, the concept of something very small, etc ...

Therefore behind every concept there is a sensation X which, if it is directly visualized (becoming the following thought) we realize that it is a mixture of other concepts and qualia. A concept is made up of a manifest qualia P, and a non-manifest or not very manifest qualia X.

Another mental phenomenon that works like that of concepts is that of music: songs, melodies, etc. Also in music there is a double aspect: the P of the sound and the X of the emotion it arouses in the listener. The difference between a music and a concept is that the X of music manages to go to the foreground, while the X of a concept is more difficult to get out of the background.

113

The same goes for art. An artistic work is a P, while the sensations and emotions it arouses are its X. Similarly there is a double aspect also in poems, myths, and, as we will see later, also in any experience.

The concept, compared to music and art, has the particularity that its X is more in the background and from here derives the sensation of its greater abstraction. Warning: the sensation of greater abstraction is indeed a sensation, a qualia, and as such it exists only and exclusively in awareness, and not somewhere out there. For this reason the feeling of objectivity, that the object exists outside awareness, is not a proof of its objectivity.

To learn a concept you need examples that allow you to form a specific X. When a person masters a certain field of knowledge well (from mathematics to carpentry), it is because she has formed good Xs: which have a lower degree of backgroundness and greater coherence. The X is a sensation that mixes several other sensations in a more or less indistinct way, without any of them standing out. X is therefore a qualia.

When we have a confused idea of something, what happens is that its X is very large and composed of a series of qualia that are not well integrated into one sufficiently distinct qualia. When, having thought about something, we have the feeling that we have clarified our ideas, what has happened is that the underlying X has broken down into a series of smaller and less nuanced Xs that are well connected to each other. A bit like tidying up a library by grouping books (read: ideas) by coherent and contiguous themes.

Therefore a concept is a form of qualia, and as such is not external to consciousness. X is what P points to and refers to. If X were not a qualia then we might assume that it is outside of consciousness. But since it is a qualia then the concepts are internal to consciousness. The consequence of this is that any conceptual effort to "get out" of consciousness to touch an external world is doomed to failure. Any conceptual architecture that can be imagined is ultimately a "qualiaplex" of P which refers to a "qualiaplex" of X. Everything resolves itself into

14. Argument from Concepts

qualia, that is, manifestations of consciousness: everything is consciousness.

Let's take an example, given P="consciousness is an emerging phenomenon in the brain", its corresponding X is a set of qualia and other concepts such as: the concept of consciousness (which is not consciousness), the concepts of phenomenon and brain, and a whole series of related images, sensations, memories and thoughts. Each of these is ultimately a qualia: "it always has a qualitative character". We never come into contact with anything that does not have a qualitative character. Behind the conceptual objects we find the qualia, and (as demonstrated in the argument from quality) behind the qualia we find only consciousness.

Concepts do not represent anything external to consciousness. Rather some represent something external to the mind. Try as we might, we are unable to get out of consciousness. But then what is the difference between the concept of an existing thing and that of a non-existent thing. For example between horse and unicorn? The horse has an X that contains images of horses perceived by the senses (e.g. eyes) and testimonials from other people who have seen horses. Instead, the X of the unicorn lacks these qualia. That's all. With respect to the concept of the physical universe horses exist while unicorns do not exist.

In Sanskrit the aspect P is called "nama", the name; while the X counterpart is called "rupa", the form. If we only consider the names, forgetting about the forms, we would have a conceptual structure: a network that connects the concepts/names/P to each other, like a vocabulary. But is this structure/vocabulary outside of consciousness? Of course not: the structure is itself a concept with an aspect P and an aspect X.

15.

THE MEANING OF EXPERIENCES

Happiness is like a butterfly:
The more you chase it,
The more it will escape,
But if you turn your attention
To other things,
It will come and sit softly on your shoulder.

Thoreau

Confused by thoughts, we experience duality in life.
Free from ideas, the enlightened see Reality.

Huang Po (Chan/Zen Buddhism Master)

When one becomes aware that both the conditioning and the apparent person who has been conditioned are ideas arising in the unconditioned being (in the Self), the dream ends, and silence, beauty and peace are revealed to oneself. This is a direct experience, it doesn't take time!

Mooji

Let us now make an excursus on the phenomenon of the double aspect in experiences, this will be useful to us later in the chapter on meaning. The double aspect, P and X, which is found in concepts, in music and in art is also present in any other experience, and it is what explains the passage from one experience to another, just as it explains the passage from one thought to another.

15. The Meaning of Experiences

It is immediate to realize that the experiences we have are in continuous flux, in continuous transformation. Heraclitus used the expression "Panta Rei", everything flows. At one moment we perceive flowers in the garden, and the next moment someone calls us from inside the house. In the becoming of experiences we also experience their continuity. We have the feeling that there is a common thread, which can be another specific experience or at least the feeling of the passage of time.

We, more or less, continuously experience experiences that express themselves through other experiences. When this does not happen then we have the feeling of a lack of meaning in the lived experiences. For example: we experience being hungry, and then the experience of starting to cook takes place. We experience a feeling of friendship towards a person and then it happens that we experience embracing that person. We experience a feeling of need to reflect, and then it happens that we find ourselves experiencing a relaxing walk in the park. In short, in our direct experience we experience that an experience E: arises, develops, and disappears as a consequence of another experience S. S is the meaning underlying E: that in dependence on which E exists (manifests itself).

Every experience E refers back to another experience S of which it is an expression, a manifestation. E is the P aspect, while S is the X aspect. Awareness is the ultimate meaning of all experience because it is wherein all other experience arises, develops and disappears.

The experience E of our dream avatar is often a manifestation of the larger experiences S of our waking mind. If before going sleep I felt a sense of anxiety S, then it could happen that during the dream I have the experience, E, of a tiger chasing me. The tiger in E is the expression of the sense of anxiety S. Our finite waking experience is an E with respect to a greater S experience: what in the Advaita tradition is called Turiya. The different experiences of different people, animals, plants, etc… are expressions of larger experiences. The author of a novel knows the experiences of his characters, but the characters do not know the experience of the author. There is asymmetry between the experience of the author, S, and the experiences of the characters, E.

We could say that two experiences "touch" each other when, at a deeper level, they are both expressions of a single greater experience. The experiences of two characters interacting (touching) in a novel are part of the reader's reading experience of the novel, or the author's one. The experiences of two people interacting are each individually an expression of the mind of the individual people; and at a deeper level they are both expressions of an experience underlying both. A first person, A, will see the second person, B, express his inner experience through a body. A experiences B's facial expressions, while the the latter is the expressions of B's inner feelings. In order for A and B's experiences to interact, they must be an expression of a single underlying experience. Like two characters in a novel, they are an expression of the mind of the author of the novel. A person is a center of meaning, a center of experience, that is, the meaning behind a set of other people's experiences. If we have an experience for which we do not feel that we are the meaning of it then another person must be behind that experience.

A dream is what a feeling or emotion looks like from within. Before falling asleep I feel a certain state of mind, a certain emotion. Then during sleep this emotion is expressed from within.

Similarly, a novel expresses a particular qualia from within. In the sense that in the novel, as well as in the dream, it is as if we try to enter the emotion by portraying ourselves as an avatar in a world in which this emotion is expressed, is unfolded, is manifested. The waking physical universe is seeing from within (or from without in the perspective of the avatar) the emotion, the qualia, of which it is an expression.

The E gives an indication of the type and depth of S. If a novel is profound, then it is safe to assume that the author's mind is profound. This observation allows us to guess that the sentience that exists in an organism is of a different category from that which perhaps could exist in a robot with AI. In fact, in the first case the S is able to express an

15. The Meaning of Experiences

organism that has generated itself. Once the egg has been fertilized, the organism develops by growing from within, and is not assembled from the outside. This S will be an experience of unity, of individuality, it will be the experience of "being a person". On the contrary, having been assembled, the robot has not built itself, and therefore its pseudo-self cannot be an experience of individuality, it cannot feel that it is a person. If you wanted to recreate a sentient being artificially, you wouldn't have to create an artificial intelligence; on the contrary, an artificial life should be created. Intelligence and sentience are distinct phenomena. The sentience would be obtained only with an artificial life. But herein lies the irony. A life can never be artificial (i.e. externally assembled) as, by definition, it generates itself.

Advaita doesn't mean "one" but it means "not-two". If there were perfect symmetry between consciousness and experiences, if there was a single experience, then it would be right to call this metaphysics as "one": enology. But since there is asymmetry then it is more correct to call this metaphysics "not-two".

I feel free when the life I live, i.e. the experiences I have, let me feel that they are an expression of my personality, of my ego, of my self. When I have an experience that I feel as no expression of myself, as a mind or a person, then I intuitively understand that I am witnessing the expression of another entity.

When our avatar in the dream meets someone or something in which he does not feel he is expressing himself, then that someone will be an expression of the dreamer's mind or perhaps of another mind... In lucid dreaming it happens that the avatar recognizes the entire dream as its expression and can decide its plot; this is because in lucid dreaming the dreamer's ego is the same as the waking's ego.

When I meet another person in waking life, I understand that their behaviors are not an expression of my mind, and therefore I understand that on the other side of that person's appearance there is another mind. Nature is not an expression of my mind, therefore I understand that on the other side of nature there is another experience-mind; perhaps that of the author of the waking dream.

When we feel a particular moment of identification with another person or natural phenomenon, it is because our mind and the one on the other side of the other person or natural phenomenon have similar expressions.

16.

THE MULTIPLICITY OF EXPERIENCES

It is neither long nor short, neither large nor small, because it transcends all limits, measures, names, traces and comparisons.

Huang Po (Chan/Zen Buddhism Master)

All there is in an experience is the knowledge I have of it.
All experiences of thinking, feeling and perceiving are a coloring of awareness, a modulation of who I am.

Rupert Spira

The drop cannot contain the ocean, but the ocean and the drop are the same water. Personal awareness cannot contain (Universal) Awareness, but it is possible to recognize that everything has One Taste. So, for the time being, you appear as a being in time, but What you really are is the Consciousness from which all this manifestation springs.

Leo Hartong

How to understand or account or explain the fact that there seem to be several organisms (and not just one) with consciousness?

Likewise, how to understand the fact that we have different experiences at different times?

These two questions can be rephrased as: at this moment another person's mind, or my mind in the past or in the future, is not accessible to me: why is it so?

In turn, this question can be rephrased as: if everything appears in consciousness then why is it that I cannot access all manifestations (past, present and future) simultaneously here and now?

By definition, a manifestation is an experience. So the question can be answered with the fact that: experiences are separate from each other! If two experiences were not separate, then they would be one experience. In fact, if two experiences were experienced simultaneously then they would be a single experience and there would be no separation between them. The question therefore translates into the fact that: there is more than "one" experience.

The separation of experiences from one another entails, "from within an experience", an apparent separation of consciousness into many consciousnesses. But in reality, it is only the experiences between them that are separate. In fact, consciousness has no separations in itself. While the experiences are many. In the argument from asymmetry, we established that objects in consciousness are separate from each other. The ego of each of us is also numbered among the objects in consciousness. The ego, that is, the one who apparently perceives the experience, is in itself a further object of experience. And as such it is separate from other egos. Because of this, a person cannot normally read another person's mind.

It is legitimate to ask why experiences are multiple? Multiplicity is real only when seen from within an experience, but not when seen from consciousness. The number one, the number two, or any other number, singularity and multiplicity, and any other mathematical entity are concepts that appear in consciousness. Consciousness is beyond numbers. Advaita does not mean "one", it means "non-two", because consciousness is beyond the singular and the plural. The main equation of non-duality is that "all is one". This equation is often read in a single direction: from "all to one". This verse exemplifies the negative path, neti-neti. We forget that the equation can and must also be read in the other direction: "one is all". This direction exemplifies the positive path. After death do we continue to exist as apparent individual

16. The Multiplicity of Experiences

consciousnesses or do we disappear into the single indistinct universal awareness? Non-dual understanding reveals to us that these are not two opposing options: they are both correct, consciousness is non-dual; in the qualitative field, unlike the analytical-quantitative one, experiences add up intensively.

Instead, why does it seem that experiences are structured within a temporal reference? Again, beware: time itself is an (content of) experience, experienced within other experiences. Every experience is in itself out of time, we could say that it is eternal: because it has no external references, all references (temporal, spatial, etc.) are within the experiences.

Suppose we fall asleep and dream that we have a degree in cosmology and are finishing a doctoral thesis on the origin of the universe in which our dream avatar lives. Suppose our avatar has collected a lot of data from a space telescope and has elaborated a mathematical model which foresees 2 possibilities. The first possibility is that the dreamed universe originated in some sort of Big Bang and that it is one of infinitely many other universes. The second possibility is that the dreamed universe has always existed. To understand which of the 2 models is correct, it would be necessary to establish the value of a cosmological constant. Unfortunately, however, it has not yet been understood how to calculate that value. While our avatar is busy figuring out how to do this calculation we wake up. Once awake, we think back to the dream and ask ourselves where and when the dreamed universe existed. With respect to the dream state, the waking state is analogous to the absolute point of view of consciousness. From this point of view, the dreamed universe has never been created, because time, space and any other reference dimension of that dreamed universe have meaning only from within that dreamed universe. For the waking state it does not matter which cosmological model would have been correct in the dream.

The origin of the dreamed universe, that is, what lies beyond it, cannot be established on the basis of the physics of the dreamed universe, that is, on the basis of the internal references of this universe. The same consideration, therefore, applies to the waking universe in which we apparently find ourselves now, as we are reading this book. This has been elaborated in the argument from simulation.

The key point is that there is asymmetry between consciousness and experiences; in the previous arguments we have in fact established that:
- Consciousness is separate from experiences.
- Experiences are not separate from consciousness, all experiences appear in consciousness, all experiences are separate from each others.

17.

THE TRIANGULATION OF EXPERIENCES

There is only consciousness and not a speck of anything else to hold onto....

Huang Po (Chan/Zen Buddhism Master)

Think for yourself, or others will think for you without thinking about you.

Henry David Thoreau

A reductionist theory of consciousness implies that consciousness is what it is described by a certain concept C, that is, by what we have called the aspect X of the concept C expressed by the word P. But X is a qualiaplex: that is, it is an object in consciousness, and therefore cannot be consciousness itself.

On the contrary when, for example, we have a certain abstract scientific theory T on a certain physical phenomenon F, what we have is that:
1. T is a set of concepts Ct, behind which there are a series of Xt.
2. F is a physical phenomenon composed of a series of sensations, images, perceptions, etc...: Xf
3. Xf is all the more a subset of Xt the more correct the theory T is.

The functioning of science is based on abstraction, which consists in the fact that the qualities Xt (numbers, mathematical structures, qualities, sensations, etc...) can be common within several other qualities Xf. This is what a scientific theory consists of: a set of objects in consciousness that correspond to other objects in consciousness. An object is never found outside consciousness. As objects in consciousness they are not consciousness. If consciousness were within the mind, and mind within the body, then it could not know the mind and the body. The objects known by consciousness are within consciousness, and not vice versa.

But then do physical objects exist or not? What happens, for example, with the following observations:
- I leave a room where the clock strikes 12:00 and I expect that after 20 minutes, as I return to the room, the clock will strike 12:20.
- Through suitable calculations I predict that the universe was born from a big bang about 13 billion years ago.
- I predict that while I sleep, people on the other side of the earth are awake and working.
- I predict that the objects present before I turn around will continue to exist behind me.

In short, by making an appropriate "triangulation" between the objects I experience, I realize that there are many objects outside my mind. How to combine this observation with the thesis of this book on non-duality?

We need to define what it is meant by "existence". An object can have independent (satyam) or dependent (mithya) existence. Well, these physical objects are mithya: they do not have their being per se and in themselves, they exist in a dependent way on other objects, and ultimately in dependence on consciousness. Physical objects certainly exist and are real, but only in relation to other objects, they do not exist in an absolute sense. In this regard, we note that:
1. The only way to discover the existence of physical objects is through appropriate inferences based on relationships with other objects. Otherwise their existence could never be established..

17. The Triangulation of Experiences

2. Physical objects "know" each other by interacting with each other. A smoke sensor knows the presence of smoke, if the smoke interacts with the sensor's sensitive membrane. If this interaction did not occur, the sensor would say that the smoke is not present: that it does not exist. If a physical object A does not interact with a second physical object B, then as regards A it follows that B does not exist.

My body and mind also have a dependent existence. On the contrary, my awareness exists in and of itself, so I know that I exist not based on other objects, but I know it in and of itself: awareness is satyam.

When I turn away from the table, the table continues to exist in relation to the other objects in the room, and in relation to my body. But the table does not exist in itself, it does not exist in an absolute sense: if the table did neti-neti it would end up finding nothing (see the argument from substantiality). In itself there is only consciousness.

An entity A exists for an entity B if A manifests itself to B. If smoke manifests itself at the sensor then smoke exists for the sensor. If an entity does not manifest itself to any other entity then we say that it "exists unmanifestly": for example, with respect to present manifestation, future events exist unmanifestly. But where is this unmanifested existence to be found? Obviously in consciousness. This is a theme that has been dealt with in the topic of becoming. Anything has always existed and will always exist in consciousness, while with respect to other things it can be manifest or unmanifest. Manifest or unmanifest existence is not in reference to consciousness, but in reference to other things/entities/manifestations (we consider these three terms as synonymous). But if I am consciousness then why am I not conscious of all manifestation? The answer was given in the chapter on the multiplicity of experiences. Here it is good to underline or remember that the mind (defined as the set of perceptions, sensations, thoughts, emotions) as well as space and time are some of the possible infinite manifestations in and of consciousness. Consciousness does not exist within mind, time and space, but it is mind, time and space that exist within consciousness. The mind, in turn, manifests itself in time and space: in the sense that the set of perceptions and thoughts that make up the mind develop within space-time coordinates. Hence the fact that a single mind is only one of infinite possible manifestations of consciousness. This is why people are not omniscient.

The fact that if two people look at the same object then they see the same thing is not explained by the fact that the two people share the outside world, but by the fact that the two people share the same consciousness in which the outside world is immersed.

As for time, if we investigate it by trying to find it we will find that we cannot find it. If I do the neti-net of time and therefore ask myself:

Where is the past? I can't find it, I only find a memory that I perceive now. I am unable to think of the past in the past. I can only think about the past now. So I exclude the past from the essential content of time.

1. Where is the future? I can't find it, I only find an expectation that I perceive now. I am unable to think about the future in the future. I can only think about the future now. So I exclude the future from the essential content of time.

2. Without the past and the future, time loses all content and therefore loses substance, ceases to be real. Time is then mithya: it exists only in dependence on something else, but not in itself.

How to make intuitive the fact that although everything is in consciousness, yet I am unable to predict everything that will happen in the universe between now and the next instant?

All our dreams are generated in our mind and proceed according to a cause-and-effect principle internal to the dream, but this does not mean that in the dream we know what will (apparently) happen in the next instant.

At this point in the book there may remain a slight uncertainty regarding the level of reality of physical objects, such as the table on which we eat, the Andromeda galaxy, a proton, my cat, etc... Are they real or not? It would seem that the materialist paradigm assigns to physical objects a somewhat greater status of reality than non-dualism can assign them. However, if we think about it, we discover that it's the

17. The Triangulation of Experiences

exact opposite. Indeed, in modern computational materialism a physical object is just a name without a form. In fact everything would boil down to a computation in a super computer. Because everything is reducible to an algorithmic structure on a substrate, and as such it is computable in a computer. Everything is just a set of names (variables, functions, relations) with no underlying reality, no further meaning, i.e. no form and substance to refer to. From this lack of form or meaning derives the consequent nihilism of the modern computational mechanism. For the latter, reality is a set of meaningless signifiers.

On the contrary, for non-dual understanding reality is consciousness manifesting itself as nama-rupa: names and forms. It's not just names, but it's also and above all forms: quality, meaning. Reality seen from a non-dual perspective is therefore much thicker than the modern computational materialist perspective sees it. Reality seen in the non-dual perspective is the most "real" that can be established: because it is founded on meanings/forms/qualities. It is therefore not a conceptual castle in the air. Reality is grounded on the basis of being, consciousness. In this regard, reference is made to the argument from quality.

To gain some understanding of what has been said, it is always helpful to use the world of dreams as an example. Suppose we are dreaming of finding ourselves sitting inside a room looking at the wall in front of our avatar. Now let's ask ourselves: does the opposite wall behind our avatar exist?

From the relative point of view of our avatar, both the wall in front of him and the wall behind him exist in the same way. Our avatar is certain that if he turns he can see the wall behind him. From the relative point of view of our ego, once we wake up, neither of the two walls has ever existed. Probably if we really wanted to give a status of existence to one of the two walls, we would be more inclined to believe that only the wall in front of our avatar existed. In fact, being outside the perceived space, where could the wall behind us have been? Nowhere, so it never existed. And what about from the absolute point of view of consciousness? From this point of view, only consciousness exists: the substance of both walls has always been only consciousness. The wall in front of the avatar was consciousness manifested with the appearance

129

(the shape) of a wall, while the wall behind the avatar was consciousness not yet manifested.

The same goes for the physical universe of our waking state: from an absolute point of view, the objects we encounter are not made up of various types of material substances, but are instead made up of a single substance: consciousness. The lemon in front of me is consciousness that manifests itself with the appearance (name and shape) of a lemon. The fir tree that I have behind me is consciousness that does not yet manifest itself with the appearance (name and shape) of a fir tree. But obviously from the relative point of view of the waking universe both the lemon and the fir exist, and they are made up of apparently distinct substances.

18.

EPISTEMOLOGY OF DIRECT EXPERIENCE

What you are, you already are. By knowing what you are not, you are free from it and can remain in your natural state. It all happens very easily and spontaneously.

Sri Nisargadatta Maharaj

If someone were to ask us, "Are you aware?", perhaps we would take a moment and then answer, "Yes". In this pause we come into contact with the most intimate and most direct experience of ourselves, and it is from this experience that the certainty of our response comes.

Rupert Spira

When you investigate the one who cares about life in the world and find that he is a phenomenon perceived just like the world, somehow, thanks to this discovery, the Self is revealed.

Mooji

Is it correct to analyze the nature of consciousness from a first person perspective? Shouldn't we study it from the third person? In fact, in the first person one does not "see" what may be out of consciousness and what causes it.

In a nutshell: does the self-inquiry of consciousness have an epistemological value?

These questions have a hidden wrong assumption. In fact, they assume that there is something outside consciousness. But this assumption is exactly the point at issue.

The key point is to realize that both what we call a first person perspective and a third person perspective are BOTH within consciousness. The first person is inside the mind and inside consciousness. The third person is outside the mind but inside consciousness.

Therefore from the point of view of consciousness (as opposed to that of the mind) there is no difference between the first and the third person perspective. Consciousness is beyond perspectives. All perspectives are within consciousness.

Therefore, it is epistemologically correct to analyze consciousness from the point of view of consciousness: because it is the same point of view with which in science we analyze physical phenomena in the third person. Therefore when, for example, paying attention to consciousness we do not see it made up of distinct parts, it is because consciousness actually has no distinct parts. The idea of stepping out of consciousness to see hypothetical parts is itself an appearance in consciousness.

The perspective with which physical phenomena are observed is the same with which one investigates consciousness: in both cases we find ourselves inside consciousness. The perspective, the point of view, the horizon, the space from which and in which we investigate physical phenomena is the same with which we investigate consciousness, and it is consciousness itself. Therefore there is no epistemological difference between physics, biology, etc... and "self inquiry".

Wanting to explain consciousness starting from physical phenomena is equivalent to explaining a painting according to the painted scene, or the ingredients of the cake according to the cake. It is the cake that is made from flour, not the other way around. Physical phenomena appear in consciousness, including their concepts, and what those concepts refer to.

18. Epistemology of Direct Experience

There cannot be an unimagined image, an unheard sound, an untasted taste, an unsavored flavour, etc... One cannot get out of consciousness and compare the image with the imagined. Because the image is made of awareness; there is nothing external to awareness in the image. Awareness is the only ingredient, substance, of which the image is made.

Similarly a thought is made of awareness. Thus an entity such as a quark, which is discovered on the basis of conceptual constructions, is made of awareness. All the sequence of reasoning, mathematical calculations, and measurements that lead us to assume the existence of quarks, and the quarks themselves all appear in awareness and their unique ingredient is always and only awareness. Quarks exist, exactly as our theories predict them, but not separately from awareness, rather separately from our mind.

The ingredients that make up a quark are the mathematical equations that describe it, the causal network in which it is inserted (the other subatomic particles), the measurements that can be made. However, these 3 intermediate ingredients are in turn made from a single basic ingredient: awareness. Because in our direct experience we see that each of these ingredients (and any ingredients of the ingredients) are not separate from awareness. The only place they manifest is in awareness. Equations are just manifestations of awareness, as are measurement graphs and the conceptual framework of particle theory.

Quarks, like all objects, are only thoughts, i.e. concepts. Concepts are physically objective to the extent that they are consistent with the sensations and perceptions we have. Beyond this coherence there is no other valid criterion of physical objectivity. The criterion of correspondence between concept and object is precluded by the fact that the object itself is not accessible, because it does not exist separate from consciousness. A concept, as well as a sensation and a perception exists only in consciousness.

From a metaphysical point of view there is agnosticism about whether or not quantum mechanics can have a connection with our mind. In this regard only physics can perhaps establish something. From a metaphysical point of view, however, it is established that everything (quantum mechanics included) is not separate from awareness.

To acquire a little insight into this type of reasoning, it is always useful to resort to the phenomenon of dreaming. Thinking that my waking brain produces consciousness is no different than thinking (while asleep) that my dream avatar's brain produces dream consciousness. Instead in both cases it is the brain in consciousness.

In a more abstract way we could formulate this reasoning as follows. To explain an entity A we have to observe everything that is observed when observing A. By observing any entity A (concrete or abstract), we end up observing at the same time a whole series of other entities Z. At this point the explanation of A is reduced to explaining the individual Zs. The explanation/observation process then has to be iterated for the individual Zs. A hypothetical entity W that is not part of the Zs cannot be added arbitrarily in this iterative process: because it is not the explanation of anything. By observing A I discover the existence of the Z, but not of the W. If it is not part of the Z, the W is to all intents and purposes to be considered non-existent. When we observe any object A (whether concrete or abstract) that object must be in awareness. But not only that: all of his Zs must also be in awareness. Therefore awareness is the final Z of everything, and we cannot assume there are any W's outside it.

19.

THE DIRECT EXPERIENCE: THE NEGATIVE PATH

Stop imagining that you are or do this or that and the realization that you are the source and heart of everything will dawn upon you.

Sri Nisargadatta Maharaj

Enlightenment is the experience of our true nature, made possible by the profound understanding of what we are not.

Francis Lucille

Can you be seen, you who perceive? If I could ask you just one question, I'd ask you this and I wouldn't stop asking you. It's such a powerful question that even if you forget everything you've heard or studied before, it will lead you home.

Mooji

Having argued against the materialist thesis and in favor of non-duality, we now have to elaborate on what worldview follows. The arguments in favor of non-dualism developed in the previous chapters, if well understood, should already give a clear idea of the vision of the world that follows. In any case, it is appropriate here to provide at least an overview of the horizon that opens up.

The most appropriate way to accomplish this task is to frame the arguments in the light of direct experience. We have to move from the analytical to the experiential realm. That is, being able to recognize in our direct experience what we discovered intellectually.

In the introduction we stated that the intellect is the bouncer of the heart. That, for analytical people it is necessary, before putting their heart, to have an approval test done by the intellect. This examination has now been done, and for anyone who has hesitations I suggest that you re-read and reflect on the previous chapters before continuing with the reading.

Human beings have elaborated, throughout their cultural history, a whole series of conceptual creations that have allowed them that they can, or made to believe of, dominate nature. Starting from the scientific revolution onwards, the most effective and efficient ways were discovered in order to create conceptual creations that would allow us to coordinate us better and bend nature to our will. We then have conceptual creations such as money, insurance, the modern state, international trade, banks, schools, universities, scientific and technological research, health care, road communications networks, internet, satellites, household appliances, permanent work, professional work, employee work, sport, volunteering, family reunions, various types of social rituals, fashion, trends, social media, real or presumed social responsibilities, common implicit beliefs, social conventions, etc…

In the contemporary world we live increasingly submerged in a network of conceptual creations that we take for granted, and we no longer even realize their origin, that we are no longer able to put into perspective and that often become ends in themselves. It's like a series of computer wires that end up tangled together and we no longer know which wire connects which things.

All these conceptual creations are like a gravitational system. In a gravitational system, such as our solar system, each planet is in a precise orbit thanks to the fact that the other planets are in their precise orbits. If one planet shifts its orbit just a little, then all the others will also shift

19. The Direct Experience: The Negative Path

their orbit appropriately until a new equilibrium orbital configuration is found. Each concept is like a planet that is in equilibrium with all the other concepts we have in mind. There's a new trend on social media, and our ideas have to change to suit us. We create ghosts to chase, and we end up being chased by them.

Everything seems to float in space without a universal inertial point of reference. This is the condition described by the theses of postmodernism: there is no absolute point of reference, everything is convention, everything is opinion. One day a research from some university comes out which informs us that they are absolutely sure that eating a certain type of food is toxic. The next day a new research comes out that says the exact opposite. One year a dream-team of virologists assures us that a certain virus is a zoonosis; and anyone who thinks otherwise is in bad faith and should be ridiculed. Few years later, the message that spreads is the opposite. There was a time when research on the link between smoking and lung cancer was funded only by cigarette manufacturers, and the result was that smoking does not cause cancer. In the end, when independent researchers have studied the phenomenon, something opposite has been discovered.

One has the impression of floating in the ocean and being pushed everywhere by the waves without being able to put one's feet on the solid bottom. Above all, one gets the impression that certain knowledge cannot be given. But in reality there is a particular area where we are not in the ocean but in a shallow pool, and if we try to stretch our legs we will find that we can touch the bottom and walk. In every solar system there is a point that acts as a permanent center of gravity and that does not move: the sun. The sun of the gravitational system of our knowledge is the direct experience. Direct experience is looking at reality as it is before superimposing our conceptual constructions on it. The latter can be considered correct only to the extent that they are consistent with direct experience. If a theory is inconsistent and contradicts direct experience then the theory is wrong. This is the great intuition of the experimental method: a theoretical hypothesis is formulated and then nature is asked if it is correct. Nature only answers questions posed in the form of an experiment. In an experiment one compares a particular direct experience with what the theory expects direct experience to be. The more the comparison shows an agreement between direct experience and theory, the more this theory is valid. Where it is not possible or impossible to compare a theory with direct experience, then the theory is worthless.

Consider the theory that consciousness is an illusion or does not exist. Our direct experience tells us that we are conscious, in fact by definition being conscious means having experiences. Therefore to say that consciousness is an illusion or does not exist contradicts direct experience and it is therefore worthless.

If we investigate what consciousness is (the Self, the ultimate subjectivity), and we strictly adhere to what direct experience shows us about it, we do not run the risk that one day a new psychological, neurological, physical or quantum theory will arrive and contradict us. Because it is the latter that on the contrary, in order to be considered valid, must not contradict direct experience. There is no escape from that.

The investigation with the direct experience into what consciousness is, it is called in Sanskrit "jnana yoga", or yoga of knowledge. The term "jnana" has the same origin as the Greek term "gnosis": which we could freely translate as knowledge. Therefore jnana yoga is the yoga of knowledge.

The jnana starts from the question of self inquiry: "Who am I?", and proceeds by exclusion, in a process called "neti-neti": not this and not that. We could see it as an inward-facing path, which slowly excludes the various contents of experience to arrive at the ultimate inner subject.

Let's look at the apple in front of our eyes and ask ourselves: am I that apple? Obviously not because next to the apple I perceive myself as the observer of the apple. On the table, next to the apple, there is a pear and I perceive that apple and pear are two distinct things because they are next to each other. Just as I perceive that the pear is next to the apple, so I perceive my Self, my I, next to the apple. My Self is therefore

19. The Direct Experience: The Negative Path

distinct from the apple: I am not the apple, I am what observes the apple.

I listen to the birds singing and at the same time the singing of an opera choir on the radio. I perceive the sounds of the two songs next to each other as well as my Self in relation to them. So I am not the sounds I hear. Similarly for any object that I perceive, be it visual, auditory, olfactory, gustatory, etc …, or a combination. I am nothing of what I perceive.

The curious thing is that the same analysis also applies to what we perceive under our skin. My body is a set of sensations of different types and intensities. Well, I perceive myself alongside these perceptions, and therefore I am not one of them or the set of all of them. Same thing for my mind. The mind is commonly understood as the aggregate of thoughts and emotions: what we call "inner experience". When I have a thought or feel an emotion at the same time I always perceive myself next to them. Just as I distinguish one thought from another because I perceive them next to each other, so I distinguish my Self from a thought.

I am always beside or in front of every perception, sensation, emotion and thought, and therefore I am distinct from them. Every perception, sensation, emotion and thought is an object because it is distinct and in front of me. But then what am I? Where am I? What do I look like? If I imagine myself with a certain aspect, I'm deceiving myself: because that image appears next to me and therefore it's not me.

Visually I can imagine being a spectator watching a show at the cinema. All my perceptions, sensations, emotions and thoughts are projected on the cinema screen. Now I will try to turn my gaze to the opposite side of the screen to find out what I see.

But I immediately notice a problem: I can't turn around, actually it's not really possible to turn around. Out of metaphor what this experiment shows us it is that I am nothing describable. I can't describe myself in terms of images (visual, auditory, olfactory, gustatory, etc...),

139

nor of sensations, nor of emotions, nor of thoughts or concepts. Any construction, even conceptual, is unable to describe me.

Let the reader try this neti-neti experiment several times, eventually trying to turn towards herself. The more she does it, the more likely she will acquire the clear awareness of not being able to describe herself.

Consciousness is indescribable. It cannot be attributed to any quality, quantity, characteristic, modality, etc … Therefore, one does not have a starting point to be able to explain it in terms of other entities. Because in order to explain an entity X in terms of another entity Y, it is first necessary to be able to characterize X, and then to show how these characteristics refer to Y.

Direct experience therefore tells us that there is no description capable of describing us. There is nothing external to me that can describe me. If there can be no external (distinct, next to) description of me then there can be no external explanation to me of me. An explanation of an entity X is a description in terms of images and concepts of the entity X. Direct experience tells us that consciousness cannot be explained by its contents. Then it must have the explanation in itself.

Therefore a neurological theory of consciousness, or a theory of consciousness as an emergent phenomenon in a physical substrate, or the theory of integrated information, cannot be considered valid descriptions and explanations of consciousness.

Consciousness is the non-objective presence of the ultimate subject, and it is what remains once all possible objects, concrete and abstract, are excluded.

This elusive aspect of consciousness has been well elaborated in the Taoist tradition. The first sentence of the TaoTeChing, the fundamental book of Taoism reads: "the Tao that can be spoken of is not the eternal Tao; the names that can be named are not eternal

19. The Direct Experience: The Negative Path

names". The Tao is consciousness: if you can describe a thing then it is not consciousness, if you cannot describe a thing then it is consciousness. There is another beautiful passage by Lao Tzu: "Only when you follow it do you lose it. But you can't get rid of it either. You can do neither, keep silent and it speaks, Speak, and it is gone. The great door of compassion is wide open and unobstructed."

In some traditions within Buddhism and Hinduism it is chosen to speak of consciousness as a "not-thing". The term "thing" is reserved for what is in some way describable. Therefore all objects that appear to our consciousness, for example: all physical entities and mental images, are things. Consciousness having no descriptions is the "emptiness", the void.

While we can't turn around to observe consciousness, what we can do is immerse ourselves in consciousness. It is a matter of placing oneself in the perspective of the witness of the experiences, without being involved in them. It is what is called meditation. Meditation is thus an investigation tool of direct experience and can be used to discover more and more aspects of it.

Putting ourselves in the perspective of the witness, what we find is that the characteristics we attribute to physical objects do not apply to consciousness. If consciousness is a void then it does not have characteristics such as "limitedness": we do not observe boundaries, edges or parts in consciousness.

But not only that: if consciousness is a void then there is nothing personal about it. The character traits of a person, his memories, his history, his cognitive abilities, etc… are all part of the mind, and appear to the consciousness, but cannot be said to be characteristics of the consciousness. From this observation we draw a first consequence: we are not our strengths and weaknesses. Strengths and weaknesses are part of our person, the self, but consciousness, our true Self, is impersonal.

The person has strengths and weaknesses, and must therefore be praised for his strengths and blamed for his defects; but not consciousness. How many times do we realize that we have made a mistake and want to correct the consequences and understand that deep down we are not that bad person. That certainly the bad person who made a mistake exists and must be punished, but that we "deep down" are not that person.

How many times does it happen that we have done something well and realize that deep down our Self has no merits. That we simply found ourselves with the right skill, intelligence or fortunate circumstances that allowed us to obtain a good result. Ultimately we do not live our life, but we are lived by life.

A second consequence of the fact that consciousness has no characteristics and attributes is that between my consciousness and yours there is no distinguishing feature or aspect. There must then be a single consciousness that "expresses" itself in a multiple number of people, human and otherwise. What is the love and compassion one feels for another person? It is the intuition that deep down we are the same consciousness: that there is unity between us.

A third consequence of the fact that consciousness has no characteristics and attributes is that it cannot be modified or changed by the contents of experience. Whatever comedy or tragedy appears as a film on the cinema screen, it does not change the cinema screen. In fact I feel that I am the same consciousness as when I was 4 years old. What I experience and my personality varies throughout life, but my consciousness always remains the same.

The experiential, not only intellectual, discovery of the non-identity of the self with the body-mind is at the basis of the Eastern idea for achieving inner peace. There is a beautiful passage from the Ashtavakra Gita, one of the oldest and most emblematic texts of Advaita Vedanta, which expresses this idea very clearly and directly: "If you detach yourself from identification with the body-mind and remain relaxed in Impersonal Consciousness and as an Impersonal Consciousness, at this very moment you will be happy, at peace, free from bondage".

19. The Direct Experience: The Negative Path

The non-dual experience, or nirvana or samadi or moksha, or whatever name you want to call it according to tradition, does not eliminate joys and pains. Joys and pains will always be there, what is transcended is suffering, understood as meta-pain: the psychological pain of having pain. And this happens because we stop identifying the Self, the true I, the consciousness, what we are, with the body, the mind, the ego, the personal history. It is not difficult to realize that the person, the ego, will always have problems to deal with. It is useless to delude ourselves that this is not the case, or that in the near or distant future it could be otherwise. On the contrary, our real Self, awareness, always remains untouched by the film projected on its screen, and is therefore happy and at peace within itself.

20.

THE DIRECT EXPERIENCE: THE POSITIVE PATH

What's wrong with this moment, unless you think about it? There's an instantaneous pause while you're trying to look to see what's wrong. Realize that before anything can be wrong, there must be a thought. Go back to that constantly. Right here, right now, currently, there's nothing wrong unless you think so.

Sailor Bob Adamson

All that we see or seem is but a dream within a dream

Edgard Allan Poe

Emotions are only the manifestation of universal being. Everything, not just the emotions, but every action, every thought, every movement, everything is included in this wonderful expression of being. Advaita is not so much concerned with interpreting any particular movement. Little attention is paid, little prominence, to the play of waves. Space is left for emotions to express themselves, exhaust themselves, and then find peace in being.

Mooji

In fluid mechanics there are two different approaches to describe a flow and consequently to write the equations that model it. In the so-

20. The Direct Experience: The Positive Path

called Lagrangian approach we ideally straddle a fluid particle and follow it along the current. In the Eulerian approach we stand on the bridge over the river and observe the whole river flowing. The Lagrangian approach is the common one when we are not in meditation, we perceive ourselves as an individual separate from the rest of the other individuals and entities with our own goals and preferences. The Eulerian approach is what we acquire during meditation or with the neti-neti. We perceive ourselves as witnesses of the whole flow of life, including our person.

The Eulerian and Lagrangian approaches are not superior to each other, but they are complementary. In fact, each provides a point of view and a perspective that the other is unable to provide. Now, however, what happens if we descend from Euler's bridge and try to put ourselves in the perspective of the river bed to perceive the entire river? This is what happens when, once the witness perspective of experiences has been established, we move forward in the investigation of experience, moving towards non-dual experience.

In non-dual experience we reconsider the objects of our experience in order to understand where they are in relation to our consciousness. Having established that we are not the objects of our experience, we ask ourselves what relationship we have with them. I look at the table in front of me where there are a pear and an apple. I perceive the apple next to the pear, and the pear next to the apple. Then I consider my consciousness and I notice that consciousness is next to the apple, but the apple is not next to consciousness: there is an asymmetry. Consciousness is separate from the apple, but the apple is not separate from consciousness. In fact: where is the apple? Is it next to consciousness? Can I find a point, real or conceptual, that separates the apple from consciousness? Can I place myself or anything else between the apple and consciousness? I cannot. There is no boundary between the apple and consciousness, the apple is totally immersed in consciousness. Everything, absolutely everything: presentations and representations, things and concepts, always manifests themself "here and now", in the single point: in awareness.

The same is true for any other experience. All experience is immersed in consciousness. There is no distance between my experience and me. It's like the distance between the landscape painted on a canvas and the canvas. Within the landscape perspective, there will be objects such as trees, rivers, animals with relative distances between them. However, trees, rivers and animals have no distance from the canvas of the picture.

Consciousness having no characteristics, being without descriptions, then it has no dimensions: there are no distances, there is no before and after, here and there, above and below. The apple is on the table because there is a three-dimensional space which gives it a mutual position. Consciousness is like a dimensionless space, in which the apple is immersed simultaneously "in all directions and from all points".

In my direct experience of the apple the only thing I find is the perception of the apple. Perception that can be composed of the image when I look at it, the touch when I grab it, the taste when I bite it, the sensation when I swallow it. If I wish, I can also add the outputs of scientific instruments that measure some properties of the apple: temperature, surface humidity, electrical conduction, weight, etc… These outputs are also my perception. If I remove perception, I remove the apple. If the apple were anything other than my perception, then I should be able to have direct knowledge of it. I have a theory that explains why an apple is on the table and considers issues such as the evolution of vegetables, the trading system, agriculture, the greengrocer etc… But all these entities and the theory that connects them are also all immersed within my consciousness in the same form in which the apple is. The image I have of the apple and all the ideas and meta-ideas about the real apple are all within my awareness. I never encounter anything outside my awareness.

In my direct experience I find nothing outside my consciousness. It's not that the apple is inside, but some quarks of some of its seeds are outside. No, everything is inside. But if nothing is outside my consciousness then it is made of consciousness, because the substance of which my perceptions, sensations, thoughts and emotions are made is

20. The Direct Experience: The Positive Path

also inside consciousness. But if everything is made of consciousness then there is nothing outside consciousness. My experiences are manifestations of consciousness in consciousness. As demonstrated in the argument from asymmetry: consciousness is transparent, consciousness does not observe an external world, on the contrary: we are always observing within ourselves.

In direct experience observer and observed are different names to speak of the same experience: observing. In the same way, giving and receiving (or selling and buying) are two different verbs to talk about the same action depending on the point of view from which it is considered. Observer, observed and observation are the same phenomenon in consciousness. Conceptually they are 3 different things, but in themselves, experientially, they are a single phenomenon. Sound and hearing are just 2 different names to indicate a single experience: hearing, which appears in and as consciousness.

In Advaita Vedanta there is an emblematic expression that encompasses everything that has been tried to be explained in this book: "atmani atmanam atmana", which can be expressed as: "knowing the Self in the Self by means of the Self". There is always and only experience that experiences experience.

Pears and apples are not separate from consciousness, but still separate from each other. This means that of them, as well as of any other object, we can and must be able to continue to speak and investigate with our best sciences, techniques and arts.

The physical universe, as well as the universes we dream of at night, are found within consciousness. If our avatar were asked in a dream where his consciousness is located, he could answer that it is inside his avatar body. However, we will know well that it is not found inside the dream body of the avatar, but rather that the entire dreamed world is inside the consciousness of the avatar, which is also the consciousness of the dreamer. Similarly, if we were to ask where our consciousness is in this physical universe, we would be making a mistake.

147

It is the physical universe that is within consciousness. Dying is therefore similar to waking up from a dream.

By carefully following what direct experience shows us, completely setting aside and without letting ourselves be distracted and influenced by theories or conceptual predispositions, we discover that everything that exists appears to consciousness, in consciousness and is made of consciousness. There is a fundamental unity of all things.

Since all things are made of consciousness, it follows that when we know anything whatsoever, we come to know consciousness itself at the same time. Hence the awareness of being aware that we feel in every experience.

Consciousness, our Self (Atman in Sanskrit), is the source of all manifestation. As a source it is what in theological terms is defined as the divine principle (the Brahman in Sanskrit). Atman is Brahman.

There is a famous Zen saying attributed to Ching-Yuan which expresses, in a peculiar Zen manner, the two steps of direct experience, the negative path of neti-neti and the subsequent positive path: "in the beginning the mountains were mountains and the waters were waters, when I penetrated into Zen wisdom, the mountains were no longer mountains and the waters were no longer waters, but when I reached the essence of Zen, the mountains were mountains again and the waters were waters again".

Both before and after the recognition of the non-duality of experience, mountains and rivers appear as mountains and rivers. Only that before they seemed to be made of something, of a substance, different from ourselves; while afterwards their fundamental essence is recognized as identical with our Self.

In the privileged moments of our life when we intuit, at least in part, this fundamental unity we experience the feeling of love and the sensation of beauty. Love is the intuition of unity with another person. Beauty is the intuition of unity with a (beautiful) object. Love is the

20. The Direct Experience: The Positive Path

dissolution of the separation between me and another person. Beauty is the dissolution of the separation between me and a thing.

With the negative path we discover that consciousness is nothing objective. But then we would be tempted to ask whether this means that nothing matters? If I'm not my own story, then who cares what I do. What does it matter to help those in need, if these are not their body-mind? The positive path shows us that objects are not separate from awareness, they are a manifestation of awareness, they are our own manifestation. Therefore the manifestation has an intrinsic value. Non-duality is also and above all to be understood as the non-duality of duality and non-duality. Therefore, the apparent duality of the manifestation has value in the deepest sense possible: it must be welcomed and accepted. Even the non-acceptance (duality) of suffering is to be accepted (non-duality). Non-duality, the Self, encloses, welcomes, loves, accepts and manifests duality. The more our "self" (our personality) imitates the "Self" in welcoming duality, the more the "self" will be at peace.

We have often wondered why mathematics is so capable of describing nature. The dilemma is that it has always been understood that mathematics is clearly a mental construction, while nature is a material system. How can something mental describe so well something that is not mental? In light of all that has been established in this book, this dilemma evaporates, it is transcended. Now we can say that mathematics has this ability due to the fact that both mathematics and physical nature are both immaterial in nature.

The exploration of direct experience provided in these last two chapters is just the tip of the iceberg. As the discerning reader will have guessed, there would still be a lot to write. However, this book is meant to be only a brief introduction to non-dualism, presented in the form that would had best satisfied my questions when I began to investigate it. For the rest, the reader should now have enough tools to continue her own investigation of direct experience on her own, possibly by reading some of the books suggested in the bibliography.

21.

THE EXPERIENCE OF MEANING

Foolish is he who hopes that our intellect,
May I reach the end of that endless road.
Canto III of Purgatory, Divine Comedy, Dante

All of existence as we know it, caged in the limitation of time, is just a reflection of that hidden principle that is continually inviting us to remember what we really are. In this reflection there is no right or wrong, better or worse, just an invitation.
Tony Parsons

Thought and in particular reasoning is like the Virgil of Dante's Divine Comedy: it can only accompany us up to a certain point of our journey. After that it must leave us in the hands of another guide. The new guide will accompany us along the road that has no end.

Don't fall into what I call the metal detector fallacy. The great success of metal detectors in detecting metals is not proof that there are only metal objects. Thus, the great success of reason is not proof that there are only things manageable by reason.

Once we have reached the limit where reason can lead us, we need to seek and find a meaning and let ourselves be led by it. There are so many aspects and choices in a person's life that cannot be established on the basis of reasoning. Only searching and following meanings (qualia) can guide us.

21. The Experience of Meaning

In every experience we recognize a moment in which reasoning and words stop and we only have to live the experience. After trying to convince someone about the goodness of mango with words, all that remains to do is to have the experience of eating the mango. The latter cannot be expressed in words. Words, speeches, are like a finger pointing to the moon. They suggest where to look, but then it's up to each of us to move our gaze from the tip of our finger to the moon, and live the experience.

If we don't take our eyes off the tip of our finger, we remain closed in a house of mirrors and get lost in them. We lose the meaning of the finger pointing at the moon, which is the moon itself. Concepts, sooner or later, must lead us to direct experience, to their ultimate meaning.

If you don't make the leap from your finger to the moon, or from the word mango to tasting a mango, you inevitably end up thinking that there is no meaning in reality. Meaning becomes a social or personal convention that has no counterpart in objective reality. Reality becomes inert, cold and mechanical. At that point, if we ask ourselves who we are, we just have to answer: "we are nothing but X", where X can be replaced with terms such as: "a chemical spit in a peripheral planet, of a peripheral galaxy", or "evil monkeys", "machines", "hamsters in a wheel", "a potential resource for the market", etc …

Everything becomes flat and gray. Certainly we can do intellectual stunts and convince ourselves that art, feelings, generosity, beauty, etc… exist, albeit only, in some way, in our mind. But that "out there" these things have no citizenship. The magic of meaning begins and ends in the short span of our lives. At most, it can linger a little longer in the memory of loved ones. But in the end it comes to the inevitable thermal, physical and spiritual death. We just have to enjoy life as long as the lights are on. Possibly sacrificing the ecosystem, and human thickness for a shred of more light.

But, at this point in the book we know quite well that we are not mere biological machines. We know that the universe and the mind are in consciousness. The mind-matter dualism was a bogus problem, mathematicians would say it was an "ill posed problem". In the sense that it was due to the fact of having assumed, perhaps without realizing it, wrong starting hypotheses without justification.

Mind and matter are both manifestations of a single substance: consciousness.

The extraordinary ability of mathematics (a purely mental construction) to describe the physical world derives from the fact that there is no duality between mind and matter. There is continuity between the two. What we call matter, the physical world, is exactly what it appears to be according to direct experience: a kaleidoscope of colours, sounds, smells, tastes, but also many other qualities that our limited senses preclude us to know. Who knows what it must be like to perceive ultrasound like bats, or electric currents like certain marine animals.

But in addition to sensible qualities, there are also many other qualia such as emotions and sensations and who knows how much more that we cannot even imagine: because a qualia is either directly experienced or it is precluded from us.

Matter, the physical world, is then just an aspect of a wider qualitative reality. Therefore matter is an expression of qualities, of which the mathematical ones are only a subset. The Romantic poets were right when they taught us to see the meaning behind any aspect of the physical world.

Physical entities such as mountains and rivers are expressions of an underlying meaning, they have a moon to point to. I see there's a moon/meaning moving my gaze from my finger/signifier to the moon. The finger image disappears, leaving room for the image of the moon.

21. The Experience of Meaning

The meaning of an experience resides in that from which it simultaneously originates, resides and finally dissolves. In other words: the meaning of an experience resides in what it is an expression/manifestation of.

There is one aspect of qualia that one need to appreciate. And it's the fact that qualia have the ability to "sum up intensively". An extensive sum is one in which the addends are placed side by side, remaining distinct and separate. The sum of 3 pears is always 3 pears next to each other. The sum of 2 qualia, X and Y, is a new qualia, Z, in which X and Y do not disappear, they are not lost, but are also no longer separate. They are rather interpenetrated, and their interpenetration is not the simple sum of X and Y, it has a further quality, it is a Z. Qualia are not solids, they are soluble fluids that mix up and can no longer be separated.

A group of notes played separately are distinct qualia, but if we play them together, we will have an intensive sum, in which the "single" notes continue to exist within a new qualia: a symphony. The notes themselves, once wrapped up in the symphony, seem transfigured because we perceive them interpenetrated.

Let's consider sugar, flour, eggs and milk: each individually has a very specific flavour: its own qualia. Now let's combine them properly, following a certain recipe, and make a cake. The cake will have a new flavor, its own qualia, distinct from that of the individual ingredients. But in any case somehow in the flavor of the cake we "recognize" the tastes of its individual ingredients in a transfigured way. The flavor of the individual components is not lost, it is only transfigured. Nothing is really lost, but everything is transfigured.

A qualia, like the flavor of the cake, contains other qualia, like the flavors of the ingredients. In turn, each of the ingredients contains the flavors of their components, and so on. There are an immense number of cake recipes, and each uses the flavors of its ingredients to express itself. The meaning of going to the market to buy eggs, flour, sugar and milk lies in being able to make the cake and express or manifest its taste.

In the argument from simulation we talked about different levels of reality, in which a lower level reality was the expression of a second higher level reality. Let us refer to that type of analysis again to understand the idea of meaning. Let's decide to call internal qualia all those that include thoughts, emotions, sensations. On the other hand, let's call external qualia all those which include sensible perceptions. The internal qualia are those that we commonly associate with a mental universe (or world or reality), while the external ones are those that we commonly associate with a physical universe (or world or reality).

Internal qualia enclose external qualia within themselves; the latter may be an expression of internal qualia. However, the reverse does not happen. For this reason a color expresses a number, its gradation; but a number does not express a colour. An abstract entity such as a number is such in that it can be an expression of many other qualia.

A myth has for meaning the feeling it wants to express. For example, the myth of the Tower of Babel tells us of a civilization so proud that it ended up building a tower that reached up to the sky, understood as a place of the divine, to be able to appropriate it. The story ends badly, with the collapse of the tower and the dispersion of languages. The meaning of the entire imaginary universe of the tower lies in expressing/manifesting the danger associated with pride, hubris. The external, physical world of the tower is an expression of the internal world of the feeling of hubris.

A novel has for its meaning, the overall emotion that the author wants to convey. Let's take a romance novel. The author through the troubled events of the protagonists wants to express, for example, the feeling of jealousy. The internal qualia of jealousy, present in the author's mind, is expressed/manifested in the external qualia of the universe of the novel.

A fairy tale has for its meaning the moral it wants to convey. Let's take Pinocchio. The imaginary universe (external qualia) of Pinocchio, populated by characters such as Geppetto, the whale, the cat and the fox,

21. The Experience of Meaning

the land of toys, is an expression/manifestation of internal qualia such as Pinocchio's filial sentiment, Geppetto's paternal love, etc…

A dream means the emotional state it expresses. Suppose, for example, that we had a stressful day and consequently were in an anxious emotional state before falling asleep. Then it will happen that we dream of being chased by a lion. The lion in the dream is an expression of the feeling of anguish in waking life. We have that an external reality, the lion, expresses/manifest an internal reality, anguish.

The novel, the myth, the fable, the dream are worlds (external qualia) that find their meaning, that is, they are an expression/manifestation of internal qualia. A set of internal qualia can be deployed, expressed, manifested, in a set of external qualia. "On the other side" of the dream lion is the feeling of anguish of waking life.

Here comes the question of the correlation between the mind and the brain. The mind is related to the brain but not caused by it: correlation does not imply causation. The mind-brain correlation indicates that internal qualia (the mind) are expressed in external qualia (the brain). Or changing the perspective: "on the other side" of the brain there is the mind.

The brain is not conscious. Consciousness is not in the brain. On the contrary, all manifestation is in consciousness. After that, all that can be said is that the external qualia of the brain, that is the physical aspect of the brain, have on the other side of the internal qualia, the mind. Similarly for a stone, a plant, etc. we will have specific internal qualia on their opposite side.

While the craftsman builds an entity starting from the assembly of different components; nature seems to operate in the opposite direction: from a simple principle it causes the multiple to spring forth. An organism is born and develops from within. A car is born and developed from the outside: from assembly in the factory. Therefore the internal qualia types on the other side of a living organism will be different from those of a car or a computer. The external qualia of a living being are an expression of internal qualia capable of expressing themselves in the form of a vital drive, of an interiority.

Since myths, dreams, novels, and stories are of the same qualitative nature as the physical universe, then the latter too must be the unfolding, the expression of some internal qualia. The physical universe is anything but meaningless. On the other side of the physical universe there is a mind, understood as a set of internal qualia, of which it is the expression.

Sunrise and sunset, a lightning storm, a calm and sunny sea, a spring in the woods, an erupting volcano, a comet trail, but also a war, a conflict, a disaster, etc… are all external expressions of some internal qualia. In the same way that in the work of an artist we have the expression of his mental state. Consider the work "The Scream of Munch": a clear expression of the painter's state of mind.

Our person, our ego, our life, are expressions of a meaning: the latter includes the former; rather than living our life, it should be said that we are lived by life. The opposite of life is not death, but birth. Life has no opposites. Dying corresponds to a change of perspective, of point of view: from external to internal, one ceases to perceive the external qualia and perceives the internal qualia of which the former were only an expression. On the other side of this universe, the qualia of this life's pain will not be explained, but will be transcended, transfigured, and thereby understood.

Consciousness is that in which all manifestations, all qualia, originate, reside, and finally dissolve. Consciousness is the ultimate meaning of everything, of which everything is an expression, and in which everything is enclosed. For this reason no experience is meaningless, and no experience is ever truly lost to oblivion.

What is the purpose of a novel? That of expressing the qualia that gives it meaning. And by unfolding it he accomplishes it. There is no qualia first and then the novel: writers often testify that writing a novel is a process of self-construction of the novel itself. An unfolding of the

21. The Experience of Meaning

qualia it is expression of. The author discovers the novel while writing. In the same way each person discovers the meaning of his life only by living: he discovers it only by living. Reason, a priori, cannot analyze meaning. Meaning is the guide that takes us over once reason has stopped. For this, Dante takes his leave of Virgil and continues with Beatrice.

Every novelist witnesses that there are no scenes or characters too many: each comes from the unfolding of the book's meaning. Every life, even the most seemingly pointless from within the narrative, has its meaning from outside the narrative: it was willed on purpose. Just as there are an indefinite number of novels, dreams, etc ... So there are an indefinite number of possible meanings, of which each person and entity is an expression.

Consciousness, having nothing outside itself, has no external constraints, therefore it expresses itself freely in the manifestation. A person is an expression of his qualia, his daimon. The perseity of a qualia means that the explanation of its way of being resides in itself. Red is red because it is red, not because of green or blue or sweet or bitter. A qualia in this sense is free from external constraints. In this way a person's daimon/qualia can be said to be free. Instead from the point of view of its expression (of the person) one cannot speak of freedom: we are not capable of choosing our next thought.

Pet owners are witnesses that on the other side of their four-legged friend's physical appearance there is a mind. Their little dog's behavior is an expression of some feeling on the part of the little dog. In the same way, and each in its own way, each thing, animal, plant, mineral, or a combination of the above, is an expression of some qualia.

There are all the indications to believe that the number of possible qualia is indefinite and inexhaustible. A mystery is not something that one cannot know, but it is something that one will never stop knowing. In this sense, consciousness is the greatest mystery. Consciousness is an inextinguishable source of meanings. There is no end to the endless road. There will always be something new to discover. It is up to each of us to live to let this source express/manifest itself.

22.

CONCLUSION

The multiplicity is only apparent, in truth there is only one mind.
Erwin Schrödinger

We are made of the same substance of dreams,
And in the space and time of a dream our short life is collected.
William Shakespeare, The Tempest, Act IV, Scene I

And the end of all our exploration will be to arrive where we started from, and know the place for the first time.
TS Eliot

It often happens to everyone to read a book and remain cold, not able to understand it, or to find it of no particular interest. Then it happens, after a certain time, even after years, to pick up the same book again and find it full of interest and meaning. At the time of the first reading we were not yet "ready". However, something was still left inside us on an unconscious level, maybe just a seed. At the right moment, perhaps after years, life has placed us in front of certain situations that have led us to seek answers. At that point the forgotten seed begins to sprout, and we remember the old book, and almost without realizing it we pick it up again and read it.

I don't know if the reader who has this book in his hands has been ready or not. What is most important is that the seed has remained. Then it will sprout at the right time.

The seed that I believe this book may have transmitted is that of:
1. Questioning the false assumptions we have about the nature of reality.
2. Investigate on your own, without relying on abstract theories, but using our direct experience. And that this investigation is feasible and fruitful.
3. Understanding that mind and awareness are two distinct things, and that not recognizing this fact leads to totally confusing the nature of reality.
4. That reality is qualitative, and therefore it is wrong to interpret everything in a reductionist way.
5. That reality is a manifestation of awareness.
6. That the non-dual understanding of reality is a stimulating and meaningful vision that does justice to our deep intuition that reality must make sense.

If the reader was ready, then I hope I helped her a little on her way. There is a phrase by Antonio Machado that reads: "walker there is no path, one makes the path by walking". Every single person creates their own path by living, without a map. But every now and then it happens to find some other walker who points out a more direct and practicable path.

Here I wanted to indicate a path, which seems particularly significant to me. After which it is up to the reader to follow him or not. In this book, only the starting point of the path is indicated, but then, often, many of us need further indications along the way. Especially with regard to its practical application in life. To this end, the nondual literature is rich, varied, and enlightening. In the bibliography I have indicated some of the authors who have been particularly useful to me.

The continuation of this path is to an important extent, at least in the beginning, constituted by non-dual meditation. This consists in meditating on "who or what am I?" in the light of what has been explained in this book. This has the aim of not remaining with a simple intellectual understanding, but of making it a direct experience: of

22. Conclusion

understanding in a direct, non-conceptual, non verbal way who or what we are. If you don't try this meditation, and are satisfied with intellectual understanding alone, then, after a while, everything learned in this book will be forgotten.

As in all things, in the end the choice of a path is not a question of brain but of heart, of instinct, of meaning. Whether a path is attractive, a person feels it or doesn't feel it. But if she feels it sometimes she needs to verbalize it a little in words to become familiar with it, so as to undertake it with more conviction. Especially for people like me, with a strictly scientific training, it is necessary to receive a particular initial indication, which gives a certain intellectual satisfaction.

The words of this book I hope have given you a sense of satisfaction sufficient to want to start the journey.

Buen Camino

APPENDIX.
PHILOSOPHICAL PRINCIPLES

You never come into contact, directly or indirectly, with any object outside your awareness. The contrary belief is in itself a further object within awareness.

Alessandro Sanna

One of the tools to argue or support a thesis in philosophy is to connect the thesis to a principle. A principle is a proposition, reason, or fundamental truth of knowledge that is very difficult or impossible to deny, and from which we can derive other truths or simply guide thought. These principles are almost always not made explicit in philosophical texts, and are often taken for granted. Certainly they are self-evident principles, but it is good to make them explicit so that the reader always knows how we got from a statement A to a statement B.

Principle of noncontradiction (PNC): an entity A cannot be A and non-A in the same sense as A and simultaneously. This principle is held to be a first principle in the sense that it is not derivative from other principles and it is self-evident in itself. Anyone trying to deny this principle should do so using this principle itself, as it is a presupposed principle in any other proof. Any other demonstration or principle always presupposes this principle.

Identity Principle (PI): An entity A is identical to itself. This principle is a corollary of the PNC.

Principle of Excluded Middle (PTE): between 2 mutually "contradictory" alternatives there cannot be a third alternative. Two contradictory alternatives exhaust all available possibilities. On the other hand, between two "contrary" alternatives, multiple alternatives can be given. For example, between two colors there is a whole gradation of tone.

Appendix - Philosophical Principles

Principle of Identity of Indiscernibles (PII): 2 entities that are not distinguished (indiscernible) in any respect are the same entity.

Principle of Sufficient Reason (PRS): Everything that exists is intelligible or everything that exists has an explanation. Explanation is not synonymous with cause. A thing may not have a cause, but it must have an explanation. Every cause is an explanation, but not every explanation is a cause: a cause is a type of explanation.

Principle of Causality (PC): A thing passes from potency to act by means of something that is already in act; everything that changes, changes through something else. The principle of causality does not say that everything has a cause, but only that what changes (that is, what passes from potency to act) must have a cause. PRS is more universal than PC.

Principle of Proportional Causality (PCP): what is found in an effect must be or pre-exist "in some way" in its causes. Because if something were found in the effect that is not in its causes, the PRS and the PC would be violated: there would be an aspect in the effect that would have no explanation and would come out of nowhere. An effect cannot have an aspect that does not come from its causes.

Principle of Finality (PF): every efficient cause has an effect. An efficient cause is an entity that causes something to "change" into something else. The "something else" is the goal.

Principle of Occam's Razor or Principle of Parsimony (PRO): causes/entities must not be multiplied without necessity. Where a single cause suffices as sufficient reason for an effect, then it is not necessary to imagine further causes. Between 2 possible explanations for a phenomenon X, the simpler one is to be preferred. Only when the simplest is no longer sufficient to explain a phenomenon we are authorized to seek a less simple explanation. The PRO is connected to the PCP: only if aspects not present in the cause are encountered in an effect, then it is necessary to look for other causes. The PRO more than a logical or metaphysical principle is a methodological principle.

Alessandro Sanna   The Direct Experience

BIOGRAPHY

There are no shortcuts to all those places worth going to.

Anon

Alessandro Sanna is a mechanical engineer with a master's degree in fluid dynamics and a PhD in computational physics. Although he specialized in Computational Fluid Dynamics (CFD) and CFD simulation software, he has been deeply interested in philosophy since childhood.

He currently lives and works between Italy, Spain and Uganda.

Alessandro Sanna The Direct Experience

ACKNOWLEDGMENT

The world exists only when we think about it; creation stories are for children. In reality the world is created every moment.

Jean Klein

I would like to thank the authors who more than others have influenced and started my philosophical and spiritual research. They are: Aina S. Erice, Vito Mancuso, Bernardo Kastrup, Rupert Spira, Robin Wall Kimmerer, Mooji, Dennis Waite, Greg Goode.

I thank Maurizia, without whom I wouldn't be who I am.

I thank Aina, the true inspiration of this book, mi escritora rarita y compañera de mi vida.

Alessandro Sanna · The Direct Experience

BIBLIOGRAPHY

We are like waves in the ocean, longing to return to the ocean that we never left. A wave experiences itself as separate from the ocean and, from that place of primal misidentification, begins to seek the ocean, in a million different ways. It is seeking itself and doesn't realize it. Its longing for home is its longing for itself. This is the human condition.

Jeff Foster

It's not enough to look, you need to look with eyes that want to see.

Galileo Galilei

There are authors who have been particularly inspiring to understand the perennial philosophy of non-duality. I know these authors only through the books they have written so far, and unfortunately I have never met them in person. However, I think they deserve to be explicitly mentioned. They are: Swami Tadatmananda, Rupert Spira, Mooji, Greg Goode, Dennis Waite, Francis Lucille, Jeff Foster, Swami Chinmayananda, Swami Dayananda, Swami Sarvapriyananda, Easwaran Eknath, Sri Atmananda Krishna Menon, Sri Ramana Maharshi and Sri Nisargadatta Maharaj. The interested reader can take a look at their works.

I found tons of other interesting authors whose works I avidly read. Each of them helped me to understand some particular aspect.

Unfortunately, time is limited and therefore I will never have the opportunity to read all the literature and do justice to all the interesting authors.

As happens in every field of knowledge, I must admit that even in this I have found less valid authors, who seem to me to transmit a misunderstood and misleading message. There are even those who, not having understood it well, give a pseudo-materialist interpretation of non-duality, and perhaps use its branding to market their products. On the opposite side are authors who combine ideas of non-duality with all sorts of wildest daydreams. For a serious researcher, the remedy against these authors is, as always, to read as much as possible from different sources, especially the original ones.

The importance of a bibliography lies, among other things, in letting the reader know that the author has possibly read and evaluated the considerations that other authors have made in their books.

While reading a book I have often wondered if the author was aware of other considerations, points of view and objections made by others.

The books below are a list, incomplete, of the points of view that I have kept in mind in formulating the arguments of my book.

1. Aczel Amir, The Mystery of the Aleph, Washington Square Press, 2001.
2. Aczel Amir, Why Science Does Not Disprove God, William Morrow, 2015.
3. Adams Robert, The silence of the heart, Infinity Institute, 1997.
4. Adyashanti, The Direct Way, Sounds True, 2021.
5. Adyashanti, True Meditation, Sounds True, 2006.
6. Alexander Eben, Proof of Heaven, Simon & Schuster, 2012.
7. Antonio Marina Jose, Biografia de la humanidad, Ariel, 2015.
8. Antonio Marina Jose, Biografia de la inhumanidad, Ariel, 2021.
9. Armstrong Karen, The Case for God, Anchor, 2010.
10. Astin John, Searching for Rain in a Monsoon, Independently published, 2013.
11. Astin John, This Extraordinary Moment, Independently published, 2022.
12. Atri Deepak, A Simple Introduction to Vedantic Non-Duality, Notion Press, 2022.
13. Balsekar Ramesh, A Duet of One, Advaita Press, 1989.
14. Balsekar Ramesh, Advaita on Zen and Tao, Yogi Impressions Books, 2016.
15. Balsekar Ramesh, Consciousness Speaks, Advaita Press, 1992.
16. Balsekar Ramesh, Final Truth, Advaita Press, 2015.
17. Balsekar Ramesh, Peace and harmony in daily living, Yogi Impressions Books, 2016.

Bibliography

18. Barrow John, The Infinite Book, Vintage, 2006.
19. Bergonzi Mauro, Il sorriso segreto dell'essere, Mondadori, 2011.
20. Berkeley George, Principles of Human Knowledge, Penguin, 1988.
21. Berkeley George, Three Dialogues Between Hylas and Philonous, Penguin, 1988.
22. Blackmore Susan, Conscious: an Introduction, Routledge, 2018.
23. Blofeld John, The Zen Teachings of Huang Po: On the Transmission of Mind, Grove Press, 1994.
24. Brockman John, This Explain Everything, Harper Perennial, 2013.
25. Brockman John, This Idea Is Brilliant, Harper Perennial, 2018.
26. Brockman John, This Idea Must Die, Harper Perennial, 2015.
27. Brockman John, This Will Make You Smarter, Harper Perennial, 2012.
28. Campbell Joseph, The Hero with a Thousand Faces, New World Library, 2008.
29. Capra Fritjof, The Hidden Connections, Anchor, 2004.
30. Capra Fritjof, The Tao of Physics, Shambhala, 2010.
31. Capra Fritjof & Lisi Pier Luigi, The Systems View of Life, Cambridge University Press, 2016.
32. Carroll Sean, Consciousness and the law of physics,, 2021.
33. Carroll Sean, Something Deeply Hidden, Dutton, 2020.
34. Carroll Sean, The Big Picture, Dutton, 2017.
35. Carroll Sean, The Biggest Ideas in the Universe, Dutton, 2022.
36. Chalmers David, Reality+, Penguin, 2023.
37. Chalmers David, The Character of Consciousness, Oxford University Press, 2010.
38. Chalmers David, The Conscious Mind, Oup USA, 1997.
39. Chopra Deepak & Kafatos Menas, You are the Universe, Random, 2018.
40. Christian Brian & Griffiths Tom, Algorithms to live by, Harper, 2017.
41. Coccia Emanule, The Life of Plants, Polity, 2018.
42. Csikszentmihalyi Mihaly, Flow, Harper Perennial, 2008.
43. D'Ors Pablo, Biografia del silencio, Galaxia Gutenberg, 2020.
44. Dalai Lama, An Introduction to Buddhism, Shanbhala, 2018.
45. Dalai Lama, Searching for the Self, Wisdom Publications, 2022.
46. Davies Paul, God and the New Physics, Simon & Schuster, 1984.
47. Davies Paul, The Demon in the Machine, University of Chicago Press, 2019.
48. Davies Paul, The Mind of God, Simon & Schuster, 1993.
49. Dawkins Richard, Outgrowing God, Black Swan, 2020.
50. Dawkins Richard, River out of Eden, Basic Books, 1996.
51. Dawkins Richard, The Blind Watchmaker, W. W. Norton & Company, 2015.

52. Dawkins Richard, The Extended Phenotype, Oxford University Press, 2016.
53. Dawkins Richard, The God Delusion, Mariner Books, 2008.
54. Dawkins Richard, The Greatest Show on Earth, Black Swan, 2009.
55. Dawkins Richard, The Magic of Reality, Free Press, 2012.
56. Dawkins Richard, The Selfish Gene, Oxford University Press, 2016.
57. Dawkins Richard, Unweaving the Rainbow, Mariner Books, 2000.
58. De Mello Antony, Awareness, The Crown Publishing Group, 1990.
59. deGrasse Tyson Neil, Letters from an Astrophysicist, WH Allen, 2020.
60. Dehaene Stanislas, Consciousness and the Brain, Penguin, 2014.
61. Dennet Daniel, Breaking the Spell, Penguin, 2005.
62. Dennet Daniel, Consciousness Explained, Penguin, 1993.
63. Dennet Daniel, Darwin's Dangerous Idea, Penguin, 2005.
64. Dennet Daniel, Freedom Evolves, Penguin, 2004.
65. Dennet Daniel, From Bacteria to Bach and Back, W. W. Norton & Company, 2017.
66. Dennet Daniel, Intuition Pumps, W. W. Norton & Company, 2014.
67. Deutsch David, The Beginning of Infinity, Penguin, 2012.
68. Deutsch David, The fabric of reality, Penguin, 1998.
69. Deutsch Eliot, Advaita Vedanta, University of Hawaii Press, 1980.
70. Dilullo Angelo, Awake, SimplyAlwaysAwake, 2021.
71. Domingos Pedro, The Master Algorithm, Basic Books, 2018.
72. Eagleman David, Brain: The Story of You, Canongate Books, 2016.
73. Eagleman David, Livewired, Canongate Books, 2020.
74. Easwaran Eknath, Conquest of Mind, Nilgiri Press, 2010.
75. Easwaran Eknath, Essence of the Bhagavad Gita, Nilgiri Press, 2011.
76. Easwaran Eknath, Essence of the Dhammapada, Nilgiri Press, 2013.
77. Easwaran Eknath, Essence of the Upanishads, Nilgiri Press, 2009.
78. Easwaran Eknath, Passage Meditation, Nilgiri Press, 2016.
79. Easwaran Eknath, The Bhagavad Gita, Nilgiri Press, 2007.
80. Easwaran Eknath, The Bhagavad Gita for daily living, Nilgiri Press, 2020.
81. Easwaran Eknath, The Dhammapada, Nilgiri Press, 2007.
82. Easwaran Eknath, The Upanishads, Nilgiri Press, 2007.
83. Easwaran Eknath, Timeless wisdom, Nilgiri Press, 2007.
84. Edelman Gerald, Neural Darwinism, Oxford University Press, 1990.
85. Edelman Gerald & Tononi Giulio, A universe of consciousness, Basic Books, 2001.
86. Einstein Albert, The World As I See It, CreateSpace Independent Publishing Platform, 2014.
87. Faggin Federico, Irriducibile, Mondadori, 2022.
88. Faggin Federico, Silicon, Waterside Productions, 2021.

Bibliography

89. Feser Edward, Aquinas: A Beginner's Guide, Oneworld Publications, 2009.
90. Feser Edward, Five Proofs of the Existence of God, Ignatius Press, 2017.
91. Feser Edward, Philosophy of Mind, St. Augustine Press, 2010.
92. Feser Edward, Scholastic Metaphysics, EDITIONES SCHOLASTICAE, 2014.
93. Feser Edward, The Last Superstition, Oneworld Publications, 2010.
94. Fiorentini Gianpaolo, Advaita Vedanta, Psiche, 2010.
95. Foster Jeff, An extraordinary absence, Non Duality, 2009.
96. Foster Jeff, The Deepest Acceptance, Sounds True, 2017.
97. Foster Jeff, The Wonder of Being, Non Duality, 2010.
98. Frankl Viktor, Man's Search for Meaning, Beacon Press, 2006.
99. Garma Chang, La Dottrina Buddista della Totalità, Astrolabio Ubaldini, 1978.
100. Ghandi Mahatma, All Men are Brother, Borodino Books, 2018.
101. Goenka S.N., The Art of Living: Vipassana Meditation, HarperOne, 2009.
102. Goff Philip, Consciousness as fundamental reality, Oxford University Press, 2017.
103. Goff Philip, Galileo's error, Vintage, 2020.
104. Goldstein Joseph, Mindfulness, Sounds True, 2016.
105. Goode Greg, After Awareness, Non Duality, 2016.
106. Goode Greg, Emptiness and Joy Freedom, Non Duality, 2013.
107. Goode Greg, Real world nonduality, New Sarum Press, 2018.
108. Goode Greg, Standing as Awareness, Non Duality, 2009.
109. Goode Greg, The Direct Path, Non Duality, 2012.
110. Hagen Steve, Buddhism plain and simple, Tuttle Publishing, 2018.
111. Hagen Steve, The Grand Delusion, Wisdom Publications, 2020.
112. Harari Yuval, Homo Deus, Harper Perennial, 2018.
113. Harari Yuval, Sapiens: A Brief History of Humankind, Harper Perennial, 2018.
114. Harari Yuval, 21 Lessons for the 21st Century, Random House, 2019.
115. Harding Douglas, On Having no Head, The Shollond Trust, 2018.
116. Harris Annaka, Conscious, HarperCollins, 2019.
117. Harris Sam, Free Will, Free Press, 2012.
118. Harris Sam, Letter to a Christian Nation, Vintage, 2008.
119. Harris Sam, Lying, Four Elephants Press, 2013.
120. Harris Sam, Making Sense, Ecco, 2020.
121. Harris Sam, The End of Faith, W. W. Norton, 2005.
122. Harris Sam, The Moral Landscape, Free Press, 2011.
123. Harris Sam, Waking up, Simon & Schuster, 2015.
124. Hartong Leo, Awakening to the dream, Non Duality, 2003.
125. Hartong Leo, From Self to Self, Non Duality, 2016.

126. Hitchens Christopher, God is not Great, Twelve, 2009.
127. Hoffman Donald, The case against reality, Penguin, 2005.
128. Hofstadter Douglas, Godel Hesher Bach, Basic Books, 1999.
129. Holecek Andrew, Dream of Light, Sounds True, 2020.
130. Holecek Andrew, Dream Yoga, Sounds True, 2016.
131. Huxley Aldous, The Doors of Perception, Vintage Classics, 2004.
132. Huxley Aldous, The Perennial Philosophy, Harper Perennial, 2009.
133. James William, The Variety of Religious Experience, Penguin, 1982.
134. Johnson Steven, Where Good Ideas Come From, Riverhead Books, 2011.
135. Jung Carl Gustav, Answer to Job, Bollingen Foundation, 2010.
136. Jung Carl Gustav, Modern man in search of a soul, Martino Fine Books, 2021.
137. Jung Carl Gustav, Synchronicity, Bollingen Foundation, 2010.
138. Jung Carl Gustav, The Undiscovered Self, Bollingen Foundation, 2010.
139. Kabat-Zinn Jon, Wherever You Go There You Are, Hachette Book, 2005.
140. Kastrup Bernardo, Brief Peeks Beyond, Iff Books, 2021.
141. Kastrup Bernardo, Decoding Jung's Metaphysics, Iff Books, 2021.
142. Kastrup Bernardo, Decoding Schopenhauer's Metaphysics, Iff Books, 2020.
143. Kastrup Bernardo, Dreamed up reality, Iff Books, 2011.
144. Kastrup Bernardo, Meaning in Absurdity, Iff Books, 2012.
145. Kastrup Bernardo, More than allegory, Iff Books, 2016.
146. Kastrup Bernardo, Rationalist Spirituality, Iff Books, 2011.
147. Kastrup Bernardo, Science Ideated, Iff Books, 2021.
148. Kastrup Bernardo, The Idea of the World, Iff Books, 2019.
149. Kastrup Bernardo, Why Materialism is Baloney, Iff Books, 2014.
150. Khaneman Daniel, Noise, Hachette Book, 2021.
151. Khaneman Daniel, Thinking fast and slow, Penguin, 2011.
152. Khenpo Tsultrim Gyamtso Rinpoche, Progressive Stages of Meditation on Emptiness, CreateSpace Independent Publishing Platform, 2016.
153. Klein Jean, Be Who You Hare, New Sarum Press, 2021.
154. Klein Jean, Beyond knowledge, New Sarum Press, 2021.
155. Klein Jean, I Am, New Sarum Press, 2021.
156. Klein Jean, The Ease of Being, New Sarum Press, 2020.
157. Klein Jean, The transmission of the flame, New Sarum Press, 2020.
158. Kock Christof, The Feeling of Life Itself, The Mit Press, 2020.
159. Kripal Jeffrey, The Flip, Bellevue Literary Press, 2019.
160. Krishnamurti Jiddu, Freedom from the Known, HarperSanFrancisco, 2009.

Bibliography

161. Kyokai Bukkyo Dando, The Teaching of Buddha, Society for the Promotion of Buddhism, 2012.
162. Lakoff George and Johnson Mark, Metaphors we Live By, The University of Chicago Press, 2003.
163. Lanza Robert, Biocentrism, BenBella Books, 2010.
164. Lao Tzu, Tao Te Ching, Independently published, 2022.
165. Lawry Kalyani, Only That, Non Duality, 2010.
166. Levy John, The Nature of Man According to Vedanta, Andesite Press, 2017.
167. Long Jeffrey, Evidence for Survival of Consciousness in NDE, HarperOne, 2011.
168. Long Jeffrey, God and the afterlife, HarperOne, 2017.
169. Loy David, EcoDharma, Wisdom Pubns, 2019.
170. Loy David, Nonduality, Wisdom Pubns, 2019.
171. Lucille Francis, Eternity Now, Truespeech Productions, 2019.
172. Lucille Francis, The Perfume of Silence, Truespeech Productions, 2010.
173. Lucille Francis, Truth Love Beauty, Truespeech Productions, 2010.
174. MacCormick John, Nine Algorithms That Changed the Future, Princeton Science Library, 2020.
175. Mancuso Stefano, Brilliant Green, Island Press, 2018.
176. Marcus Gary, Kluge, Mariner Books, 2009.
177. McGilchrist Iain, The Master and His Emissary, Yale University Press, 2019.
178. McGilchrist Iain, The Matter With Things, Perspectiva, 2021.
179. Melanie Mitchel, Complexity, Oxford University Press, 2011.
180. Mooji, An Invitation to Freedom, Mooji Media Publications, 2022.
181. Mooji, Before I Am, Mooji Media Publications, 2012.
182. Mooji, Breath of the Absolute, Mooji Media Publications, 2016.
183. Mooji, Vaster Than Sky Greater Than Space, Sounds True, 2016.
184. Mooji, White Fire, Mooji Media Publications, 2020.
185. Norbu Namkhai, Dzog-Chen, Snow Lion, 2003.
186. Osborne Arthur, The Teachings of Ramana Maharshi, Rider, 2014.
187. Osho, The Mustard Seed, Osho Media International, 2009.
188. Parson Tony, The Open Secret, Open Secret Publishing, 2010.
189. Pearl Judea, The Book of Why, Basic Books, 2020.
190. Peterlini Sergio, Bhagavad Gita, Edizioni il Punto d'Incontro, 1996.
191. Peterlini Sergio, Ribhu Gita, Edizioni il Punto d'Incontro, 2018.
192. Peterlini Sergio, Sri Tripura Rahasya, Edizioni il Punto d'Incontro, 2001.
193. Peterlini Sergio, Srimad Bhagavatan, Edizioni il Punto d'Incontro, 2019.
194. Peterlini Sergio, Yoga Vasistha, Edizioni il Punto d'Incontro, 2021.
195. Peterson Jordan, 12 Rules for Life, Penguin, 2019.

196. Pievani Telmo, Imperfection, The MIT Press, 2023.
197. Pinker Steven, Enlightenment Now, Penguin, 2019.
198. Pinker Steven, How Mind Works, W. W. Norton, 2009.
199. Pinker Steven, Rationality, Viking, 2021.
200. Pinker Steven, The Better Angel of our Nature, Penguin, 2012.
201. Pinker Steven, The Blank Slate, Penguin, 2003.
202. Plotino, Enneadi.,,
203. Radhakrishnan S., Bhagavad Gita, HARPER COLLINS PUBLISHERS, 2020.
204. Ram Dass, Being, Sounds True, 2022.
205. Ramachandran V.S., The Tell-Tale Brain, W. W. Norton & Company, 2012.
206. Ramana Maharshi, Be as You Are, Penguin, 1998.
207. Ramana Maharshi, Who Am I?, Sri Ramanasramam, 2008.
208. Raphael, The Pathway of Non Duality, Aurea Vidya, 2016.
209. Reale Giovanni, Storia della Filosofia Greca e Romana, Bompiani, 2018.
210. Red Pine, The Lankavatara Sutra, Counterpoint, 2013.
211. Ridley Matt, The Evolution of Everything, Harper Perennial, 2016.
212. Ridley Matt, The Rational Optimist, Fourth Estate, 2011.
213. Ridley Matt, The Red Queen, Harper Perennial, 2012.
214. Rosling Hans, Factfulness, Flatiron Books, 2020.
215. Rovelli Carlo, Che cosa è la scienza. La rivoluzione di Anassimandro, Mondadori, 2017.
216. Rovelli Carlo, Helgoland, Riverhead Books, 2022.
217. Rovelli Carlo, Reality Is Not What It Seems, Riverhead Books, 2018.
218. Rovelli Carlo, Seven Brief Lessons on Physics, Riverhead Books, 2016.
219. Rovelli Carlo, The Order of Time, Riverhead Books, 2019.
220. Russell Peter, From Science to God, New World Library, 2004.
221. Russell Peter, Letting go of nothing, New World Library, 2021.
222. Russell Stuart, Human Compatible, Penguin, 2020.
223. Sacks Oliver, An Anthropologist On Mars, Vintage, 1996.
224. Sacks Oliver, The River of Consciousness, Vintage, 2018.
225. Sailor Bob Adamson, Presence- Awareness, Independently published, 2019.
226. Sailor Bob Adamson, What's wrong with right now, Independently published, 2019.
227. Schroedinger Erwin, Mind and Matter, Cambridge University Press, 1958.
228. Schroedinger Erwin, My View of the World, Cambridge University Press, 2009.
229. Schroedinger Erwin, Nature and the Greek and Science and Humanism, Cambridge University Press, 1996.

Bibliography

230. Schroedinger Erwin, What is Life?, Cambridge University Press, 2012.
231. Seng Tsan, Hsing Hsing Ming.,,
232. Seth Anil, Being You, Faber and Faber, 2022.
233. Shankara, Atma Bodha, Kessinger Publishing, 2010.
234. Spira Rupert, A Meditation on I Am, Sahaja, 2021.
235. Spira Rupert, Being Aware of Being Aware, Sahaja, 2017.
236. Spira Rupert, Being Myself, Sahaja, 2021.
237. Spira Rupert, Presence Volume I The Art of Peace and Happiness, Sahaja, 2016.
238. Spira Rupert, Presence Volume II The Intimacy of All Experience, Sahaja, 2016.
239. Spira Rupert, The Essential Self, Sahaja, 2023.
240. Spira Rupert, The Nature of Consciousness, Sahaja, 2017.
241. Spira Rupert, The Transparency of Things, Sahaja, 2016.
242. Spira Rupert, You are the happiness you seek, Sahaja, 2022.
243. Sri Atmananda Krishna Menon, Atma Darshan, Advaita Publishers, 1991.
244. Sri Atmananda Krishna Menon, Atma Nirvriti, Advaita Publishers, 1991.
245. Sri Atmananda Krishna Menon, Notes on Spiritual Discourses, Non Duality, 2017.
246. Sri Nisargadatta Maharaj, I Am That, Chetana Private Ltd, 1999.
247. Sri Nisargadatta Maharaj, Prior to Consciousness, Acorn Press, 1990.
248. Sri Nisargadatta Maharaj, Prior to Consciousness, Acorn Press, 1990.
249. Sri Nisargadatta Maharaj, Seeds of Consciousness, Acorn Press, 1990.
250. Sri Nisargadatta Maharaj, The Experience of Nothingness, Blue Dove Press, 1996.
251. Suzuki D.T., Essays in Zen Buddhism, Souvenir Press, 2022.
252. Suzuki D.T., The Zen Koans as a means to Attain Enlightenment, Charles E. Tuttle, 1994.
253. Suzuki D.T., Zen Buddhism: Selected Writings, Harmony, 1996.
254. Swami Chinmayananda, Aitareya Upanishad, Central Chinmaya Mission Trust, 2018.
255. Swami Chinmayananda, Ashtavakra Gita, Central Chinmaya Mission Trust, 2018.
256. Swami Chinmayananda, Atma Bodha, Central Chinmaya Mission Trust, 2012.
257. Swami Chinmayananda, Kaivalyopanishad, Central Chinmaya Mission Trust, 2012.

258. Swami Chinmayananda, Kathopanishad, Central Chinmaya Mission Trust, 2012.
259. Swami Chinmayananda, Kenopanishad, Central Chinmaya Mission Trust, 2012.
260. Swami Chinmayananda, Isavasyopanishad, Central Chinmaya Mission Trust, 2015.
261. Swami Chinmayananda, Mandukya Upanishad with Karika, Central Chinmaya Mission Trust, 2012.
262. Swami Chinmayananda, Mundakopanisad, Central Chinmaya Mission Trust, 2008.
263.
264. Swami Chinmayananda, Taittiriya Upanisad, Central Chinmaya Mission Trust, 2014.
265. Swami Chinmayananda, The Holy Geeta, Central Chinmaya Mission Trust, 2020.
266. Swami Chinmayananda, Vivekachoodamani, Central Chinmaya Mission Trust, 2006.
267. Swami Dayananda, Drg Drsya Viveka, Arsha Vidya, 2022.
268. Swami Dayananda, Introduction to Vedanta, Vision Books, 1997.
269. Swami Dayananda, Tattvabodhah, Arsha Vidya, 2012.
270. Swami Dayananda, The Teaching of the Bhagavad Gita, Vision Books, 2005.
271. Swami Nikhilananda, Drg-Drsya-Viveka, Advaita Ashrama, 1931.
272. Swami Nikhilananda, Gospel of Ramakrishna, Sri Ramakrishna, 2010.
273. Swami Sarvapriyananda, Dissolve into infinity, Juggernaut, 2021.
274. Swami Sarvapriyananda, The Atman, Juggernaut, 2021.
275. Swami Sarvapriyananda, What is Vedanta, Juggernaut, 2021.
276. Swami Sarvapriyananda, Who Am I?, Juggernaut, 2021.
277. Swami Satchidananda, The Yoga Sutra of Patanjali, Integral Yoga Publications, 2012.
278. Swami Vivekananda, Jnana Yoga, Vedanta Press, 2011.
279. Swami Vivekananda, Karma Yoga and Bhakty Yoga, Vedanta Press, 1999.
280. Tegmark Max, Our Mathematical Universe, Vintage, 2015.
281. Tegmark Max, Life 3.0, Penguin, 2018.
282. Thaler Richard H. & Sunstein Cass R., Nudge, Penguin, 2021.
283. Thich Nhat Hanh, Diamond That Cuts Through Illusion, Parallax Press, 2006.
284. Thich Nhat Hanh, The Art of Living, HarperOne, 2017.
285. Thich Nhat Hanh, The Miracle of Mindfulness, Beacon Press, 1996.
286. Thich Nhat Hanh, You are Here, Shambhala, 2010.
287. Tolle Eckhart, A New Earth, Penguin, 2008.
288. Tolle Eckhart, Oneness with all life, Penguin, 2020.

Bibliography

289. Tolle Eckhart, Practicing the Power of Now, Hodder & Stoughton, 2002.
290. Tolle Eckhart, Stillness Speaks, New English Library, 2003.
291. Tolle Eckhart, The Power of Now, Coronet, 2002.
292. Tollifsson Joan, Nothing to Grasp, Non Duality, 2012.
293. Tononi Giulio, Phi, Pantheon, 2012.
294. Trungpa Chogyam, Cutting Through Spiritual Materialism, Shanbhala, 2002.
295. Van Lommel Pim, Consciousness Beyond Life, HarperOne, 2011.
296. Waite Dennis, Answers, Mantra Books, 2020.
297. Waite Dennis, AUM, Mantra Books, 2015.
298. Waite Dennis, Back to the Truth, Mantra Books, 2017.
299. Waite Dennis, The Book of One, Mantra Books, 2010.
300. Wall Kimmerer Robin, Braiding Sweetgrass, Milkweed Editions, 2015.
301. Watson Burton, The complete work of Zhuangzi, Columbia University Press, 2013.
302. Watts Alan, Become what you are, Shambala, 2003.
303. Watts Alan, Tao, Faber and Faber, 2019.
304. Watts Alan, The Book, Souvenir Press, 2009.
305. Watts Alan, The way of zen, Rider, 2021.
306. Williams Paul, Mahayana Buddhism, Routledge, 2008.
307. Wilson Edward, Biophilia, Harvard University Press, 1984.
308. Wilson Edward, The Meaning of Human Existence, Liverright, 2015.
309. Wright Robert, Nonzero, Vintage, 2001.
310. Wright Robert, The evolution of God, Back Bay Books, 2010.
311. Wright Robert, The Moral Animal, Vintage, 1995.
312. Wright Robert, Why Buddhism is True, Simon & Schuster, 2018.
313. Yogananda Paramahansa, Autobiography of a Yogi, Self-Realization Fellowship, 2000.
314. Yogananda Paramahansa, Demystifying Patanjali The Yoga Sutras, Crystal Clarity Publishers, 2013.

Alessandro Sanna The Direct Experience

Printed in Great Britain
by Amazon